JN422608

ACING PHYSICS 11

Review of Physics 11 for Students in BC

June Chung
Patricia Grindlay

유한문화사

ACING PHYSICS 11

Review of Physics 11 for Students in B. C.

Dedicated to All the Scientifically Inquisitive Young Minds

Copyright © 2007 by June Chung

Copyright © 2007 by June Chung

All rights reserved. No part of this publication may be reproduced, stored in a retrieval system, or transmitted, in any form or by any means, electronic, mechanical, photocopying, recording or otherwise, without the prior written permission of the copyright owner.

Direct comments, suggestions, questions, and requests to:

JC Publishing
#1004 - 933 Hornby St.
Vancouver, BC V6Z 3G4
e-mail: acingphysics@hotmail.com

Acing Physics 11 - Review of Physics 11 for Students in B.C.

ISBN 0-9781249-0-2

Author: June Chung

- Ph. D., University of Illinois at Urbana-Champaign, 2002
- M.S., Carnegie Mellon University, 1997
- B.S., Carnegie Mellon University, 1996

Associate Author & Editor: Patricia Grindlay

- Head of Physics & Chemistry, Little Flower Academy, Vancouver, B.C., 1985-2006
- Ed. Diploma, University of London, England, 1970
- B.Sc. (Tech), University of Manchester, England, 1961

Acknowledgement: Adele Coady

- Head of Science, Little Flower Academy, Vancouver, B.C.

Printed in S. Korea

PREFACE

Welcome to Physics 11! Many students believe that physics is very difficult and you need to be a genius to enjoy it – an idea encouraged by countless films and TV programs featuring very intelligent-looking, white-coated scientists saving the world! In truth, physics is logical and straightforward although it does require some mathematical ability.

Physics 11 is the course that introduces you to the most fundamental physical concepts such as velocity, acceleration, force, energy, waves, etc. As is true in any other discipline, it is essential to build a strong understanding of these concepts before progressing to the more advanced courses. This book is intended to help students gain insight into the ideas introduced in this course and use it to answer questions.

To use this book effectively, you should read each Key Point and make sure you understand the concepts explained in it. Then try to connect these concepts to the example problems at the end of that Key Point. Also, there is a set of questions at the end of each chapter to test your knowledge on vocabulary, understanding of major concepts, and ability to apply those concepts. After taking the chapter test, analyze the results. If you are uncertain about any of the ideas introduced in the chapter, go back and repeat the process all over again. The explanations in this book are concise enough to allow you to review the material repeatedly.

I believe that, when used as recommended above, this book will definitely help you improve your performance in and enjoyment of Physics 11 and also lay a solid foundation for future physics courses. I would love to hear your comments or suggestions on this book, so please direct them to the contact information shown on the previous page. I wish you all good luck, and have fun!

Sincerely,

June Chung

June Chung

Patricia E. Grindlay

Patricia Grindlay

TABLE OF CONTENTS

CHAPTER 1

INTRODUCTION

Physics is the study of change and the energy which is involved in that change. Nobody knows exactly what energy is - only what it does. As we cannot see it, hold it, or measure it, we cannot examine it directly. Instead, we have to define change in terms of quantities that we can measure, and express relationships between these quantities using the language of mathematics. Then we can determine the energy associated with the measured change. As measurement is very important in physics, we need to learn how to record measurements properly and perform mathematical operations with them accurately.

In physics, there are a number of quantities that require us to measure the direction in addition to the magnitude. Examples include velocity, acceleration, and force. Such quantities are called vectors and play an important role in describing measured changes. Hence, we need to get familiar with vectors and know how to perform basic operations with them.

In this chapter, you will study:

- ✓ significant digits and scientific notation
- ✓ how to perform mathematical operations using measured quantities
- ✓ scientific units
- ✓ scalars and vectors
- ✓ graphical representation of vectors
- ✓ operations involving vectors

Copyright © by June Chung - All rights reserved. Reproduction of this publication without the written permission of the copyright owner is strictly prohibited.

Key Point 1.1: Significant Digits and Scientific Notation

significant digits

In physics we frequently deal with numbers that come from measurements of an object's attributes or its motion. No measurement can be absolutely perfect because every measuring tool we use is only accurate to a certain degree. Accordingly, when writing measurements or performing calculations with them, we should use only the reliable digits, known as **significant digits**. In this course, you should be able to recognize the number of significant digits in each measurement; perform calculations with them; and express the result with the right number of significant digits.

counting the number of significant digits

The number of significant digits is straightforward if there is no zero in a number. For instance, there are three significant digits in the measurement 175 cm, and five in 67.255 kg. Here are the rules when zeros appear in a number:

1. Zeros at the beginning of a measurement are not counted as significant. Thus, 0.0035 km has only two significant digits.
2. Zeros between non-zero digits are counted as significant. Thus, 2008 years and 0.90305 m/s have four and five significant digits, respectively.
3. Zeros at the end of a measurement may or may not be significant. It is hard to know the correct number of significant digits in this case. For example, 400 °C might have one, two, or three significant digits. However, when a measurement has a decimal point, all the zeros are counted as significant. Hence, 400.0 °C has four significant digits.

scientific notation

In order to avoid confusion about the number of significant digits and to allow the expression of very big and small numbers in a concise manner, we often write numbers and measurements using **scientific notation**. In scientific notation, a number is expressed as

$$x \times 10^n \text{ (for positive numbers)}$$

$$\text{or } -x \times 10^n \text{ (for negative numbers)}$$

where $1 \le x < 10$, and n is an integer. Examples are shown below:

$$4 \times 10^2; \; 4.0 \times 10^2; \; 4.00 \times 10^2$$

Copyright © by June Chung - All rights reserved. Reproduction of this publication without the written permission of the copyright owner is strictly prohibited.

All the numbers above are written as 400 when expressed in standard form, but it is easy to see that the number of significant digits is different in each case.

> - Every measurement involves uncertainty.
>   ⟹ Use only reliable digits (significant digits)
> - Scientific notation is used...
>   1. to clarify the number of significant digits.
>   2. to express very big and small numbers in a concise manner.

Example 1-1

Identity the number of significant digits in each number.

a) 344 b) 6020 c) 0.12080 d) 7.99001500

Example 1-2

Express the following numbers using scientific notation.

a) 8000000 (use four significant digits)

b) 0.0123

c) 574.411

d) 0.00000083990

Copyright © by June Chung – All rights reserved. Reproduction of this publication without the written permission of the copyright owner is strictly prohibited.

Key Point 1.2: Calculations Using Measured Quantities

addition of measured values

When measured values are added or subtracted, the result cannot be more precise than the least precise value. For example, when three numbers 9.10, 21.3, and 115 are added, the calculation is done as follows:

$$\begin{array}{r} 9.10 \\ 21.3? \\ +\ 115.?? \\ \hline 145.40 = 145 \end{array}$$

We round off our answer to 145 because we are unsure about the first and second decimal place. However, if the three numbers had been 9.20, 21.4, and 115, the calculation would have become

$$\begin{array}{r} 9.20 \\ 21.4? \\ +\ 115.?? \\ \hline 145.60 = 146 \end{array}$$

In this calculation, we round up our answer to 146 because we know the first decimal place is at least 6, which is greater than 5.

multiplication of measured values

In multiplication or division, the outcome should have the same number of significant digits as the measurement with the least. For instance, the following calculation is answered using one significant digit since 3 is the number with the fewest significant digits:

2 significant digits... (2.7)
1 significant digit... (3)

$$2.7 \times \frac{3}{5.00} = 1.62 = 2$$

← so, 1 significant digit!

3 significant digits... (5.00)

Be careful when you use either multiplication or conversion factors as they have <u>infinite</u> precision. For example, if you measure the radius of a circle to be 2.6 cm, the diameter is 5.2 cm, not $2.6 \times 2 = 5$ cm. The number 2 in the equation is not from a measurement. By agreement, the diameter of a circle is precisely twice its radius, and, therefore, the number 2 has an infinite number of significant digits.

a series of calculations with measured values

When performing a series of calculations involving measured values, do not round up or off during the calculation. Instead, keep more digits than necessary during the calculation, and give the answer with the

Copyright © by June Chung – All rights reserved. Reproduction of this publication without the written permission of the copyright owner is strictly prohibited.

correct number of significant digits at the very end. For example, consider a cube of side 3.2 m. When calculating the volume of the cube, you may multiply 3.2 m three times in a row:

$$V = 3.2 \times 3.2 \times 3.2 = 32.768 = 33 \text{ m}^3$$

Or, you may find the base area first, which is

$$A = 3.2 \times 3.2 = 10.24 \text{ m}^2$$

and multiply it by the height. However, if you round off the base area to 10 m^2 and then multiply it by the height, you obtain

$$V = A \times h = 10 \times 3.2 = 32 \text{ m}^3$$

which is different from the answer obtained without rounding off. Therefore, do not worry about significant digits until you reach the end. In this example, because we know the volume of the cube should have two significant digits, we adjust our answer accordingly from 32.768 m^3 to 33 m^3 at the very end.

> - In additions and subtractions, the result cannot be more precise than the least precise value.
> - In multiplications and divisions, the result must have the same number of significant digits as the measurement with the least.
> - Do not worry about significant digits during a series of calculations; give the final answer with the correct number of significant digits.

Example 1-3

Perform the following calculations and express the answers with the proper number of significant digits.

a) $3.1025 + 5.31 - 0.098$

b) $2.90 \times 8.8 \div 7$

c) $\pi \times 6.3 - 1.25$

Example 1-4

James converted 10.5 seconds, his record for a 100-m sprint, into minutes as follows. Is his calculation correct? If not, what should be the correct conversion?

3 significant digits... (10.5 sec) 1 significant digit... (1 min)

$$10.5\,\text{sec} = 10.5\,\text{sec} \times \frac{1\,\text{min}}{60\,\text{sec}} = 0.2\,\text{min}$$ ← so, 1 significant digit!

2 significant digits... (60 sec)

Copyright © by June Chung – All rights reserved. Reproduction of this publication without the written permission of the copyright owner is strictly prohibited.

Key Point 1.3: Scientific Units

importance of units

When specifying a measured value, you should always attach the proper unit. Without a unit, we cannot determine which physical quantity is being referred to, nor the magnitude of that quantity. For instance, the number 3 fails to tell us whether it refers to a length, a time, a force, or something else. Even if we knew it referred to a time period, we would not know whether it is 3 seconds, 3 days, or 3 years.

fundamental and derived units

The fundamental units used in Physics 11 are those of mass, length, time and temperature. In SI (Système International) units, they are respectively the kilogram [kg], the meter [m], the second [s], and the Kelvin [K]. All other units that you will see later in the course, such as the Newton [N], the Joule [J], and the Watt [W], are derived units as they can be expressed in terms of the fundamental units. (See Appendix B, p. 218.)

metric prefixes

In the metric system, the specification of larger and smaller units is facilitated by the use of **metric prefixes**. The ones you encounter often in this course are presented in Table 1-1. Thus, 1.23×10^6 grams can be written as 1.23 Mg, and 0.0000437 seconds can be written as 43.7 μs.

Table 1-1: Metric prefixes

Prefix	Value
giga (G)	10^9
mega (M)	10^6
kilo (k)	10^3
milli (m)	10^{-3}
micro (μ)	10^{-6}
nano (n)	10^{-9}

- Always use the proper unit with a measured value.
- The fundamental units in Physics 11 are the kilogram (kg), the meter (m), the second (s), and the Kelvin (K). All other units are derived units.
- Metric prefixes facilitate the specification of the larger and smaller units.

Example 1-5

a) Express 1.000 day in seconds, and then in Ms.

b) Express 1.58×10^{-12} m in units of nm and mm.

Copyright © by June Chung – All rights reserved. Reproduction of this publication without the written permission of the copyright owner is strictly prohibited.

Key Point 1.4: Scalars and Vectors

scalars

A **scalar** quantity has magnitude only. Examples of scalar quantities are mass, time, distance, and speed. These quantities can be expressed as a number with a unit (as in 65 kg, 60 s, 42.195 km, 25 m/s, and so on).

vectors

A **vector** quantity has both magnitude and direction. Thus, if a quantity has a direction, you should recognize it as a vector. Otherwise, it is a scalar. Examples of vector quantities are displacement, velocity, acceleration, and force. (You will learn about them in the following chapters.)

discerning vectors from scalar quantities

To decide whether a given quantity is a scalar or a vector, think about whether it has a direction. For instance, a person's age is a scalar quantity; there is no direction in which a person ages. Similarly, frequency and temperature are scalar quantities because they can be expressed as a single number without a direction. In fact, most scientific quantities you have learned so far are scalars. So, it will be easier to memorize the vector quantities introduced in this course and regard the rest as scalars.

> - A scalar quantity has magnitude only.
> - A vector quantity has both magnitude and direction.
> - If a quantity has a direction, it is a vector. Otherwise, it is a scalar.

Example 1-6

Identity each quantity as a scalar or a vector.

a) 765 kByte
b) 15 km, west
c) -15 °C
d) 30 km/h, north

Copyright © by June Chung – All rights reserved. Reproduction of this publication without the written permission of the copyright owner is strictly prohibited.

Key Point 1.5: Graphical Representation of Vectors

representation of a vector

While a scalar quantity is represented by a single letter, a vector quantity is represented by a letter with an arrow over it (for example, $\vec{V}$) or by a bold letter (**V**). Graphically, it is represented by an arrow (as shown in Fig. 1-1). The length of the arrow is proportional to the vector's magnitude. Thus, a vector with a magnitude of 30 m/s is twice as long as a vector with a magnitude of 15 m/s and three times as long as a vector with a magnitude of 10 m/s. Remember that the magnitude of a vector can never be negative.

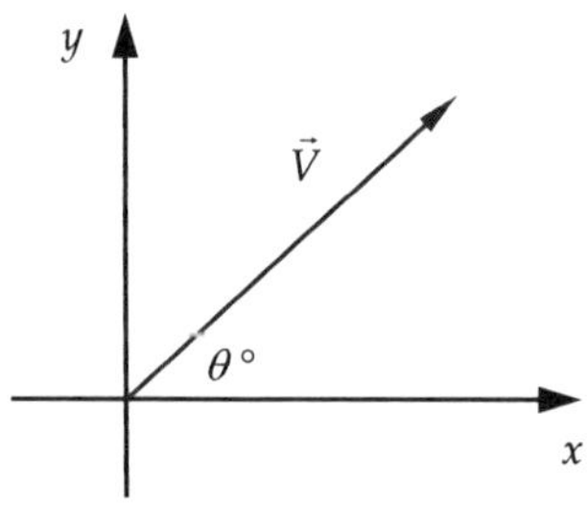

Figure 1-1: A graphical representation of a vector

specification of a vector's direction

The direction of the arrow coincides with that of the corresponding vector. To specify a direction, we usually determine a reference direction first and then measure the angle in a clockwise or counter-clockwise direction from that reference direction. For instance, you can say the vector in Fig. 1-1 has the direction of $\theta°$ above the $+x$-axis (or $(90 - \theta)°$ clockwise from the $+y$-axis).

If two vectors possess the same magnitude and direction, they are identical vectors. It means that, even if a vector is translated horizontally or vertically, it is still the same vector because its magnitude and direction remain unchanged.

> - Vectors are represented by arrows.
>   arrow's length = vector's magnitude
>   arrow's direction = vector's direction
> - Identical vectors have the same magnitude and direction.

Copyright © by June Chung – All rights reserved. Reproduction of this publication without the written permission of the copyright owner is strictly prohibited.

Key Point 1.6: Operations Involving Vectors

Vectors can be multiplied or divided by scalars, added to or subtracted from other vectors, and multiplied by other vectors[†]. However, adding vectors with scalars is not allowed.

multiplying a vector by a scalar

When a vector is multiplied or divided by a positive scalar quantity, its magnitude changes accordingly with no change in direction. Suppose you multiply a vector, say $\vec{A}$, by 4. Then the product is a vector whose magnitude is four times that of $\vec{A}$, and the direction is unchanged.

When a vector is multiplied or divided by -1, its direction is reversed with no change in magnitude. Hence, multiplying $\vec{A}$ by -4 can be interpreted as increasing its magnitude by a factor of 4 and reversing its direction.

$$\vec{A}\times(-4)=\vec{A}\times(-1)\times 4$$

multiply magnitude by 4 (the factor 4)

reverse direction (the factor (−1))

adding vectors

When two vectors are added or subtracted, the result is another vector called the **resultant vector**. You must pay attention to the vectors' directions when adding them. For example, adding 2.1 m, east, with 1.2 m, east, yields a resultant vector of 3.3 m, east. However, adding 2.1 m, east, with 1.2 m, west, produces a resultant vector of 0.9 m, east. To understand this more clearly, we can show the vector addition graphically as shown in Fig. 1-2.

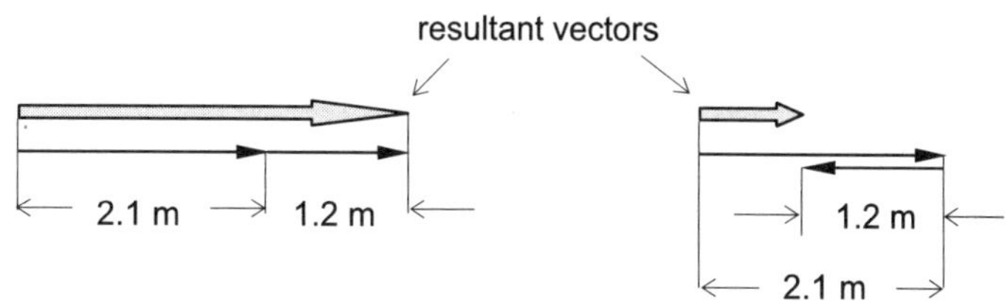

(a) vectors in same direction

(b) vectors in opposite directions

Figure 1-2: Addition of vectors

[†] Multiplication of vectors is beyond the scope of Physics 11 and will not be discussed in this book.

Copyright © by June Chung – All rights reserved. Reproduction of this publication without the written permission of the copyright owner is strictly prohibited.

tip-to-tail method

To add vectors graphically, first draw one of the vectors, and keep drawing each vector so that the second one starts from the tip of the first, the third one starts from the tip of the second, and so on. The resultant vector is then the arrow drawn from the tail of the first vector to the tip of the last one. This graphical approach is called the **tip-to-tail method**, and an example is given for an addition of four vectors in Fig. 1-3.

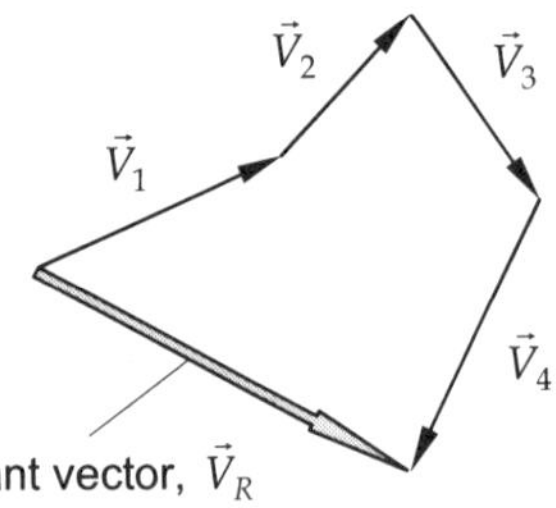

Figure 1-3: Addition of vectors using the tip-to-tail method

subtracting vectors

To subtract a vector from another vector (as shown in the equation below), first reverse its direction and then add it to the other vector. Hence, subtracting 1.2 m, west, from 2.1 m, east, is the same as adding 1.2 m, <u>east</u>, to 2.1 m, east.

$$\vec{A} - \vec{B} = \vec{A} + \left(-\vec{B}\right)$$

direction is reversed!

- When a vector is multiplied by a scalar, its magnitude changes. If the scalar is negative, the direction of the vector is reversed as well.
- When adding vectors, pay attention to their directions!
- To subtract a vector, reverse its direction and then add it.
- Use the tip-to-tail method to add vectors graphically.

Example 1-7

Given that $\vec{A} = 32$ m, north, and $\vec{B} = 12$ m, south, perform the following operations:

a) $3.0 \times \vec{A}$ b) $-2.5 \times \vec{B}$ c) $\vec{A} + \vec{B}$ d) $\vec{A} - \left(2.0 \times \vec{B}\right)$

Copyright © by June Chung – All rights reserved. Reproduction of this publication without the written permission of the copyright owner is strictly prohibited.

ANSWER KEY TO EXAMPLES

Example 1-1

a) 3 digits

b) either 3 or 4 digits (The last zero may or may not be a reliable digit.)

c) 5 digits

d) 9 digits

Example 1-2

a) 8.000×10^{6}

b) 1.23×10^{-2}

c) 5.74411×10^{2}

d) 8.3990×10^{-7}

Example 1-3

a) The answer must be precise to the second decimal place (because of 5.31). Thus, the answer is 8.31.

b) The answer should be expressed with one significant digit (because of 7). The answer is 4.

c) Since π has an infinite number of significant digits, the answer must have two significant digits because of 6.3. Hence, $\pi \times 6.3 - 1.25 = 19.79 - 1.25 = 18.54 = 19$.

Example 1-4

Conversion factors have infinite precision. Therefore, when 10.5 s is converted into minutes, the answer should have <u>three</u> significant digits.

Thus, $10.5\,\text{sec} = 10.5\,\text{sec} \times \dfrac{1\,\text{min}}{60\,\text{sec}} = 0.175\,\text{min}$

Example 1-5

a) 1.000 day = 24.00 hours = 1440 minutes = 86400 seconds

86400 s = 8.640×10^{4} s = 0.08640 Ms

b) 1.58×10^{-12} m = $1.58 \times 10^{-3} \times 10^{-9}$ m = 1.58×10^{-3} nm

1.58×10^{-12} m = $1.58 \times 10^{-9} \times 10^{-3}$ m = 1.58×10^{-9} mm

Example 1-6

(a) scalar; (b) vector; (c) scalar; (d) vector

Example 1-7

(a) $3.0 \times \vec{A} = 96$ m, north

(b) $-2.5 \times \vec{B} = -1 \times 2.5 \times \vec{B} = 30$ m, north

(c) $\vec{A} + \vec{B} = 20$ m, north

(d) $2.0 \times \vec{B} = 24$ m, south; $-\left(2.0 \times \vec{B}\right) = 24$ m, north

Hence, $\vec{A} - \left(2.0 \times \vec{B}\right) = \vec{A} + \left(-2.0 \times \vec{B}\right) = 56$ m, north

Copyright © by June Chung – All rights reserved. Reproduction of this publication without the written permission of the copyright owner is strictly prohibited.

PROBLEMS ON CHAPTER 1: INTRODUCTION

(Answer Key on p. 223)

1 – 5. Fill in the blanks

1. Since every measurement involves uncertainty, we express measured values using __________ __________.
2. The SI units of mass, length, and time are ______, ______, and ______, respectively.
3. In the metric system, the specification of the larger and smaller units is facilitated by the use of __________ __________.
4. A __________ is a quantity with magnitude only.
5. A __________ is a quantity possessing both magnitude and direction.

6 – 15. True/False questions

6. _____ A measured value of 2.100 has two significant digits.
7. _____ The number of significant digits is always clear when a measured value is expressed in scientific notation.
8. _____ 1000 nm is equivalent to 1 μm.
9. _____ The height of a building is a scalar quantity.
10. _____ The longer the length of an arrow, the greater the magnitude of the corresponding vector.
11. _____ If two vectors are parallel, they have the same direction.
12. _____ Two vectors with the same magnitude and the same direction are identical.
13. _____ If $\vec{B} = 2\vec{A}$, the magnitude of $\vec{B}$ is twice that of $\vec{A}$.
14. _____ If $\vec{B} = -\vec{A}$, the direction of $\vec{B}$ is opposite to the direction of $\vec{A}$.
15. _____ If $\vec{A} + \vec{B} = \vec{0}$, the vectors have the same magnitude and opposite directions.

Copyright © by June Chung – All rights reserved. Reproduction of this publication without the written permission of the copyright owner is strictly prohibited.

16 - 20. Multiple choice questions

16. We represent all measured values using significant digits because

    a) only the first several digits are meaningful.
    b) every measurement has an uncertainty.
    c) carrying too many digits is inefficient.
    d) we want to be able to express them in scientific notation.

17. Which of the following calculations is correct?

    a) $13.025 + 1.3025 = 14.3275$ b) $10.0 - 7 = 3.0$
    c) $5.98 \times 2 = 12$ d) $5/8 = 0.6$

18. Which of the following is equivalent to 0.01 GHz?

    a) 1×10^{10} μHz b) 1×10^{6} Hz
    c) 10,000 kHz d) 0.1 MHz

19. Which of the following quantities is a vector?

    a) speed b) time
    c) force d) mass

20. For the two vectors shown below, which of the following diagrams shows the correct method to find $\vec{A} - \vec{B}$?

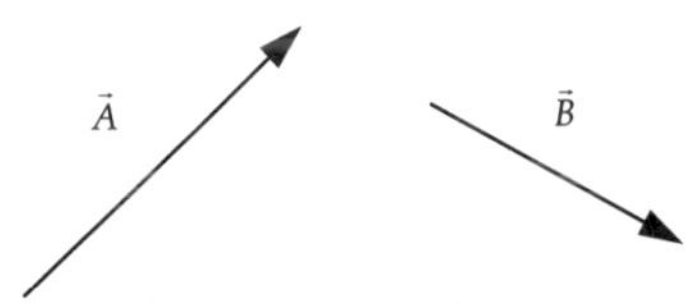

a)
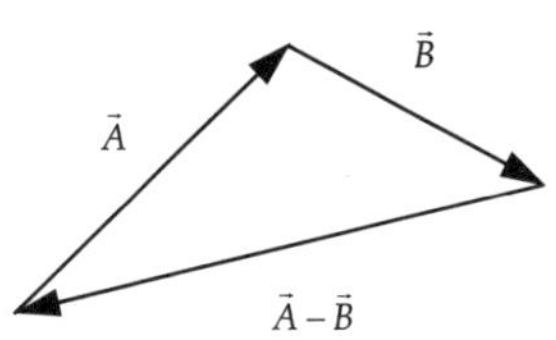

b)
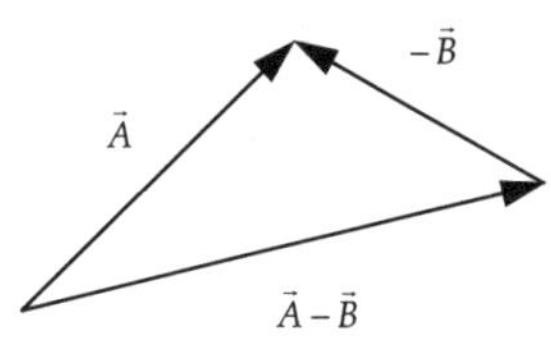

c)
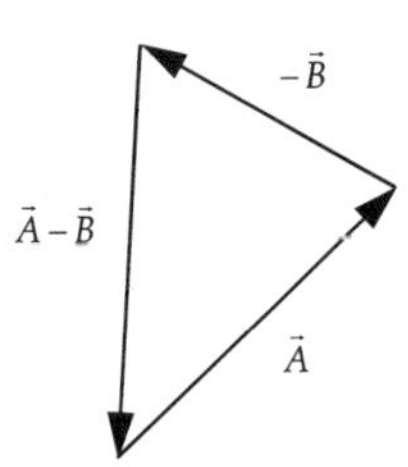

d)
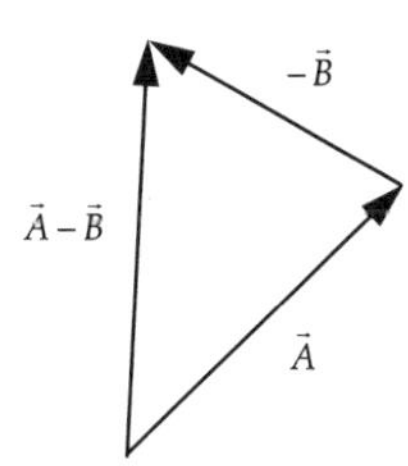

Copyright © by June Chung - All rights reserved. Reproduction of this publication without the written permission of the copyright owner is strictly prohibited.

21 - 25. Written-response questions

21. Perform the following operations on measured values and express the answers in scientific notation.

    a) 3.142 cm + 12.34 cm

    b) 12.8251 s - 4.367 s

    c) 28.3 m × 13.225 m

    d) 127 km ÷ 0.73 hr

22. Convert the metric prefixes of the following measured values.

    a) 3.51 GHz into MHz

    b) 0.29 μs into ns

    c) 670 kBytes into MByte

    d) 4.8 ng into mg

23. Given that $\vec{A}$ = 3.7 m, north, and $\vec{B}$ = 1.5 m, south, calculate $2\vec{A} - 3\vec{B}$.

24. Determine the resultant of the following four vectors: 3.2 km, north; 2.7 km, east; 5.1 km, south; and 2.7 km, west.

25. If $\vec{A}$ = 5.00 m, south, and $\vec{B}$ = 12.0 m, west, find the magnitude of their vector sum.

Copyright © by June Chung - All rights reserved. Reproduction of this publication without the written permission of the copyright owner is strictly prohibited.

CHAPTER 2

KINEMATICS

Kinematics is the study of motion, without consideration of why the object moves. We already know a lot about motion because we have been moving ourselves and watching moving objects since childhood. Experience tells us that motion involves position, time, and direction. With these three simple ideas, kinematics enables us to quantify more complex quantities like displacement, speed, velocity, and acceleration, and to understand how they are related to one another. Using these relationships we can determine how long it will take a ball thrown upward to come back to the ground, or how far a car travels while accelerating uniformly for a certain time.

In this chapter, you will study:

✓ position, displacement, velocity, and acceleration

✓ the differences between displacement and distance

✓ the differences between velocity and speed

✓ the graphical analysis of one-dimensional motion

✓ the motion of free-falling objects

✓ kinematic equations, and when you may use them

✓ how to solve kinematic problems using kinematic equations

Copyright © by June Chung – All rights reserved. Reproduction of this publication without the written permission of the copyright owner is strictly prohibited.

Key Point 2.1: Position ($\vec{x}$)

position is a vector

Position is a quantity that refers to an object's location. It is a vector, meaning you must specify how far and in what direction the object is located relative to your position. For instance, suppose someone asks you where city hall is. If you tell that person only how far it is ("It's 5.5 km away.") or only the direction ("It is north of here."), they would not be able to find it. In order to locate city hall, you need to say how far he or she needs to travel and in which direction ("It is 5.5 km north of here.").

position vector is relative

A position vector is represented graphically with an arrow, starting from the location of the observer to that of the object. Figure 2-1 shows the position vectors for objects located at A and B. Notice that the position of an object is relative to the location of the observer. For example, the positions of A and B as seen by Observer P are different from those as seen by Observer Q. It is obvious that an object located three meters directly in front of you could be five meters behind your friend and one meter to my left.

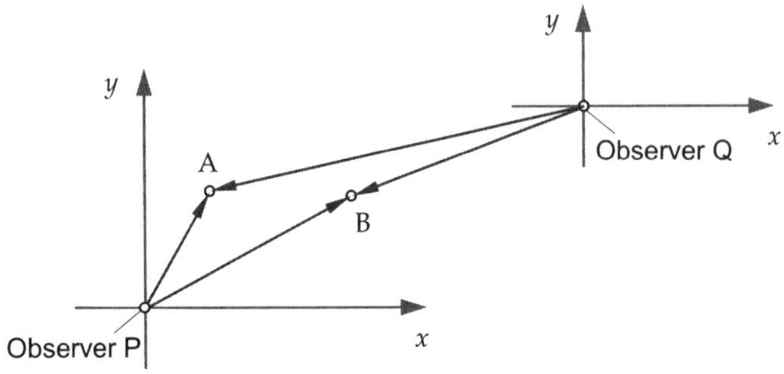

Figure 2-1: Position vectors

> - Position is a vector that refers to an object's location.
> - An object's position is relative to the observer's location.

Copyright © by June Chung – All rights reserved. Reproduction of this publication without the written permission of the copyright owner is strictly prohibited.

Key Point 2.2: Displacement ($\vec{d}$)

displacement vector

Displacement is a vector quantity that refers to a change in position. Suppose an object is initially located at A in Fig. 2-1 and then moves to a new location, B. We can represent this displacement (or change in position) graphically by drawing a straight arrow from the initial position, A, to the final position, B, as shown in Fig. 2-2. Note that, unlike position vectors, displacement vectors do not depend on the location of the observer.

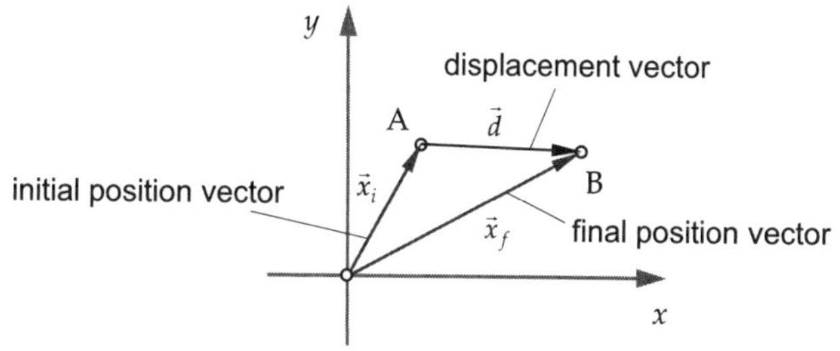

Figure 2-2: Drawing a displacement vector

relationship between position vectors and displacement

From the vector diagram in Fig. 2-2, the following relationship between the position vectors and the displacement vector is established:

$$\vec{x}_i + \vec{d} = \vec{x}_f \quad [\text{m}] \qquad \ldots (2.1)$$

We can re-write Eq. (2.1) to express displacement in terms of the initial and final positions:

$$\vec{d} = \vec{x}_f - \vec{x}_i = \Delta\vec{x} \quad [\text{m}] \qquad \ldots (2.2)$$

The Δ sign means 'change in.' Thus, if the letter t means time, Δt means 'change in time.' Likewise, $\Delta\vec{x}$ means 'change in position.' To find the change in some physical quantity, always subtract its initial value from its final value.

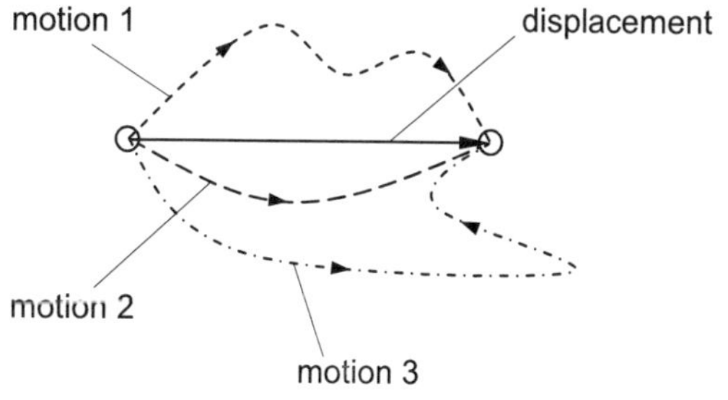

Figure 2-3: Three different motions with identical displacement

Copyright © by June Chung – All rights reserved. Reproduction of this publication without the written permission of the copyright owner is strictly prohibited.

Keep in mind that a displacement vector depends only on the initial and final positions and has nothing to do with the actual path of travel. For instance, the three different motions shown in Fig. 2-3 have the same displacement because their initial and final positions are identical.

> - Displacement is a vector that refers to an object's change in position.
> - Displacement does <u>not</u> depend on the observer's location.
> - The displacement of an object depends only on its initial and final positions and <u>not</u> on the path of travel.
> - Finding a displacement vector graphically:
>   1. Find the initial and final positions of the object.
>   2. Draw a straight arrow from the initial position to the final position.
>   3. The arrow is the displacement vector.

Example 2-1

Find the displacement of each object:

a) A soccer ball, dropped from 1.1 m above the ground, bounces a few times and comes to rest on the ground.

b) A coin, tossed vertically upwards from 0.5 m above the ground, is caught on its way down at 0.8 m above the ground.

c) An explorer travels 25 km east, 10 km north, 12 km east, 15 km south, and then 37 km west.

Copyright © by June Chung – All rights reserved. Reproduction of this publication without the written permission of the copyright owner is strictly prohibited.

Key Point 2.3: Displacement vs. Distance

distance

Distance is the length of the actual path of travel taken by an object. It is a scalar quantity, which means it has no direction. Therefore, you cannot compare distance with displacement. Note that displacement does not depend on the actual path of travel while distance does. For example, it has already been pointed out that the three motions in Fig. 2-3 all have the same displacement. However, the travel distance is different for each motion (with motion 3 traveling the farthest).

distance vs. displacement

Now, consider the following question. If you run 100 meters west in a straight line, what is your displacement and distance of travel? Are they the same? If not, which one is greater? If you have read the previous paragraph carefully, you should know that you cannot compare distance with displacement directly! In this case, the displacement is 100 m, west, and the travel distance is 100 m. Thus, the magnitude of your displacement, which is 100 m, is the same as the distance you ran.

> - Distance is the length of the actual path of travel. It is a scalar quantity.
> - Displacement is a vector and does not depend on the actual path of travel. You need only the initial and final positions to find displacement.
> - Remember to always specify the direction of the displacement!

Example 2-2

Tommy walks 1.6 km to school in the morning and comes back home after school via the same route. Find his displacement and distance of travel.

Copyright © by June Chung – All rights reserved. Reproduction of this publication without the written permission of the copyright owner is strictly prohibited.

Key Point 2.4: **Velocity ($\vec{v}$)**

Velocity is a vector quantity that indicates the rate at which an object is changing its position. There are two kinds of velocity: average velocity and instantaneous velocity. To calculate average velocity, one needs to measure the displacement (or change in position) of an object and divide it by the time of travel. In mathematical form, it is expressed as

average velocity

$$\vec{v} = \frac{\Delta \vec{x}}{\Delta t} = \frac{\vec{d}}{\Delta t} \text{ [m/s]} \qquad \ldots (2.3)$$

Equation (2.3) shows that average velocity becomes zero whenever the displacement becomes zero. Therefore, it is possible for an object to have an average velocity of zero even though it has been constantly moving. For instance, if a runner runs around a track and returns to the starting position, the average velocity will be zero no matter how fast he or she runs!

instantaneous velocity

Instantaneous velocity, as the name suggests, is the velocity at an instant in time. Thus, the runner in the above example will have a non-zero instantaneous velocity as long as he or she is moving. Note that the direction of the instantaneous velocity is always the same as the direction in which the object is moving. Accordingly, a basketball bouncing on the floor will have the instantaneous velocity vector pointing upward while rising and downward while falling.

- Velocity is a vector that indicates the rate at which an object changes its position.
- Finding an average velocity vector:
  1. Find the displacement vector.
  2. Divide the displacement by time of travel.
- Whenever displacement becomes zero, average velocity becomes zero as well.
- Instantaneous velocity is the velocity at an instant in time; its direction is always the same as that of the motion.

Copyright © by June Chung – All rights reserved. Reproduction of this publication without the written permission of the copyright owner is strictly prohibited.

Key Point 2.5: Velocity vs. Speed

speed

Speed is a scalar quantity that indicates the rate at which an object is covering the ground. You can find an object's average speed over a period of time by dividing the distance traveled by the time taken. Note that average velocity is obtained by dividing the displacement by the time taken.

> - Average speed = distance / time
> - Average velocity = displacement / time

average velocity vs. average speed

Accordingly, the magnitude of average velocity is only equal to average speed when the magnitude of displacement is the same as the distance traveled. As shown in Key Point 2.3, the magnitude of displacement is the same as the distance traveled only when an object travels in a straight line. Otherwise, the distance is always greater than the magnitude of the displacement. Thus, average speed is always greater than or equal to the magnitude of average velocity.

instantaneous velocity vs. instantaneous speed

As for the magnitude of instantaneous velocity and instantaneous speed, they are always equal to each other. For example, if an object's instantaneous velocity is 80 km/h, east, then its instantaneous speed is 80 km/h. Also, when an object becomes momentarily motionless, its instantaneous velocity and instantaneous speed both become zero at that instant.

From here on, instantaneous velocity and instantaneous speed will be referred to simply as 'velocity' and 'speed,' respectively. Problems involving average quantities will clearly state that you need to use or find average velocity or average speed.

> - Average speed is always greater than or equal to the magnitude of average velocity.
> - Instantaneous speed is always equal to the magnitude of instantaneous velocity.

Copyright © by June Chung - All rights reserved. Reproduction of this publication without the written permission of the copyright owner is strictly prohibited.

Example 2-3

A Ferris wheel with a radius of 25 m is rotating at a rate of 0.20 revolutions per minute. If Becky gets on the Ferris wheel at its lowest point, calculate (a) her displacement and distance of travel after half a revolution and (b) her average velocity and average speed during half a revolution.

Example 2-4

(a) Cathy leaves her home in Vancouver at noon and travels south for 200 km to arrive at her friend's house at 2:30 p.m. Find Cathy's average speed and average velocity.

(b) After spending a couple of days at her friend's place, Cathy is now returning to Vancouver. She leaves her friend's at noon, heading north. After having traveled for 50 km, she suddenly remembers that she left her physics book at her friend's house. So, she goes back, retrieves her book, and finally arrives home at 4:00 p.m. Find Cathy's average speed and average velocity.

Copyright © by June Chung - All rights reserved. Reproduction of this publication without the written permission of the copyright owner is strictly prohibited.

Key Point 2.6: **Acceleration ($\vec{a}$)**

acceleration

Acceleration is a vector quantity that indicates the rate at which an object is changing its velocity. It is expressed as

$$\vec{a} = \frac{\Delta\vec{v}}{\Delta t} = \frac{\vec{v}_f - \vec{v}_i}{\Delta t} \quad [\text{m/s}^2] \qquad \dots (2.4)$$

where $\vec{v}_i$ and $\vec{v}_f$ refer to the instantaneous velocities at the initial and final times, respectively. For example, if an object moving at 25 m/s, north, changes its velocity to 35 m/s, north, in 2.0 seconds, its average acceleration will be

$$\vec{a} = \frac{35-25}{2.0} = 5.0 \text{ m/s}^2\text{, north}$$

acceleration and change in velocity

According to Eq. (2.4), whenever the velocity of an object stays constant, its acceleration is zero since $\vec{v}_i = \vec{v}_f$. If the velocity is not constant, it means that there is a change in the speed or the direction of motion or both, and the object experiences a non-zero acceleration.

direction of acceleration

Keep in mind that the direction of an acceleration vector is independent of the direction of motion. For instance, you might think that an object thrown vertically upward experiences an upward acceleration while it is rising. However, the object always experiences a downward acceleration while in the air, whether it is rising or falling. The general rule of thumb is that, if an object gains speed while traveling in a straight line, its acceleration vector points in the direction of the motion. If an object loses speed, its acceleration points in the opposite direction to the motion.

> - Acceleration is a vector that indicates the rate at which the velocity changes.
> - Finding an acceleration vector:
>   1. Find the initial and final velocity vectors.
>   2. Subtract the initial velocity from the final velocity. Be careful: it is a vector subtraction!
>   3. Divide the resultant vector by time of travel.
> - The direction of the acceleration is independent of the direction of the motion.

Copyright © by June Chung – All rights reserved. Reproduction of this publication without the written permission of the copyright owner is strictly prohibited.

Example 2-5

Debra's car is initially moving north at 55 km/h. Within 15 seconds, its velocity changes to 30 km/h, south. Find the average acceleration of Debra's car. (Be careful of the units!)

Example 2-6

An object initially moving east at 25 m/s accelerates twice: first at 2.0 m/s^2, east, for 23 seconds and then at 5.1 m/s^2, west, for 11 seconds. What is the final velocity of the object?

Copyright © by June Chung – All rights reserved. Reproduction of this publication without the written permission of the copyright owner is strictly prohibited.

Key Point 2.7: Graphical Analysis of One-Dimensional Motion

one-dimensional motion

When an object moves along a straight line, we say its motion is one-dimensional. As shown in Fig. 2-4, a car traveling on level ground, a box sliding on an inclined surface, and a ball tossed vertically into the air are examples of one-dimensional motion. Each vector quantity describing such a motion (that is, displacement, velocity, and acceleration) is directed either in the same direction as the motion or in the opposition direction to it. This allows us to specify a vector with a number (indicating its magnitude) and a + or – sign (indicating its direction). Then we can plot a graph of either position ($\vec{x}$) versus time (as shown in Fig. 2-5) or velocity ($\vec{v}$) versus time (as shown in Fig. 2-6).

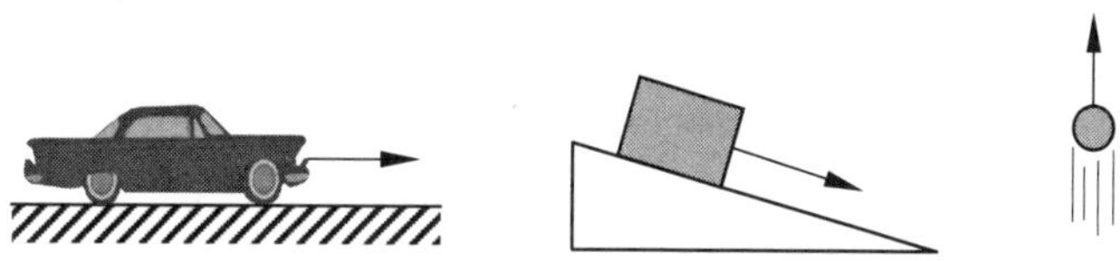

Figure 2-4: Examples of one-dimensional motions

finding displacement from a graph of position vs. time

For instance, a graph of position versus time is shown in Fig. 2-5. To find the object's displacement between t_0 and t_1, we use Eq. (2.2):

$$\vec{d} = \vec{x}_f - \vec{x}_i = +12 - 0.0 = +12 \text{ m}$$

The plus sign indicates that the object moved 12 m in the positive direction. Likewise, the displacement between t_1 and t_2 is

$$\vec{d} = \vec{x}_f - \vec{x}_i = -4.0 - (+12) = -16 \text{ m}$$

Thus, the object moved 16 m in the opposite direction. You can find the object's total displacement by adding the two displacements or by using the positions at t_0 and t_2. Either way, you will obtain the displacement to be 4 m in the negative direction.

finding average velocity from a graph of position vs. time

To find the object's average velocity between t_0 and t_1, we apply Eq. (2.3):

$$\vec{v} = \frac{\Delta\vec{x}}{\Delta t} = \frac{+12 - 0.0}{3.0 - 0.0} = +4.0 \text{ m/s}$$

This value of +4.0 m/s is exactly the slope of the line segment between t_0 and t_1. ($\Delta\vec{x} = +12.0$ m is the rise, and $\Delta t = 3.0$ s is the run.) In fact, the

Copyright © by June Chung - All rights reserved. Reproduction of this publication without the written permission of the copyright owner is strictly prohibited.

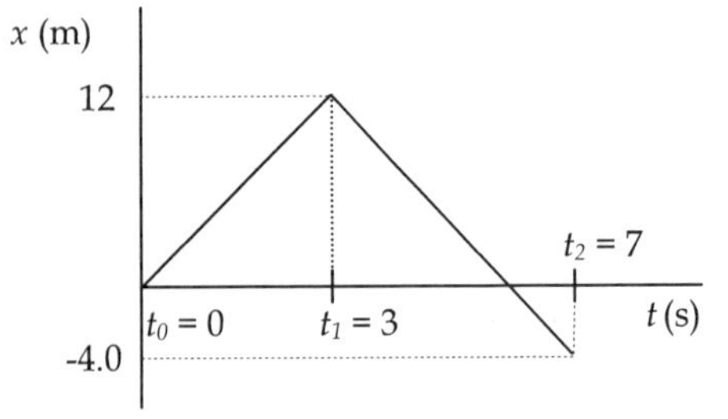

Figure 2-5: A graph of position vs. time

slope of the graph represents the velocity. The sign of the slope (that is, + or –) indicates the direction of the velocity vector, and the steepness of the slope represents the magnitude of the velocity, or speed.

The motion in Fig. 2-5 can be described as follows:

> An object initially moves in the positive direction at 4.0 m/s until it has traveled 12 m in 3.0 seconds. Then it turns back and travels in the opposite direction at 4.0 m/s for 4.0 seconds, finally arriving at a location 4.0 m on the other side of the initial position.

finding acceleration from a graph of velocity vs. time

Examine how acceleration is defined in Eq. (2.4), and it is easy to see that the slope of a velocity vs. time graph represents the acceleration. Hence, the sign of the slope indicates the direction of the acceleration vector, and the steepness represents its magnitude. The motion in Fig. 2-6 has uniform acceleration as the slope of the graph is constant.

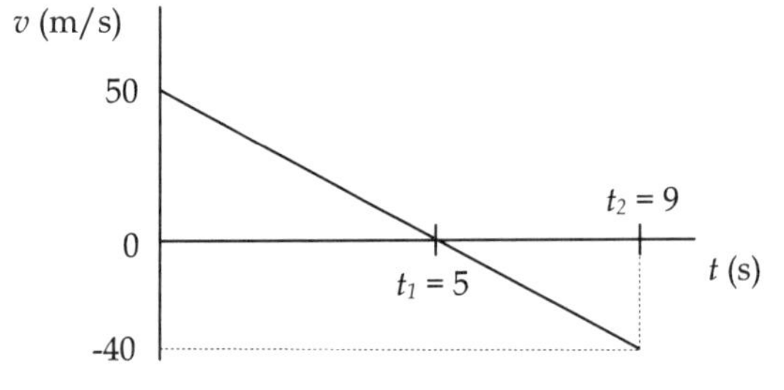

Figure 2-6: A plot of velocity vs. time

finding displacement from a graph of velocity vs. time

You can also find the displacement of an object from a $\vec{v}$ vs. t graph by calculating the area between the graph and the horizontal axis. Above the horizontal axis, the area indicates a displacement in the positive direction. Below the axis, the displacement is in the negative direction. In Fig. 2-6, the object thus travels

Copyright © by June Chung – All rights reserved. Reproduction of this publication without the written permission of the copyright owner is strictly prohibited.

$$\frac{1}{2} \times 50 \times (5.0 - 0.0) = +125 \text{ m}$$

during the first five seconds, and

$$\frac{1}{2} \times (-40) \times (9.0 - 5.0) = -80 \text{ m}$$

during the next four seconds. In total, the object's displacement is 45 m in the positive direction.

- Displacement: <u>area</u> under velocity vs. time graph
  → Area above the horizontal axis is considered positive, and below is considered negative.
- Velocity: <u>slope</u> of position vs. time graph
- Acceleration: <u>slope</u> of velocity vs. time curve
  → Sign of the slope reveals the direction; Steepness reveals the magnitude.

Example 2-7

Consider the graph of position vs. time in Fig. 2-5.

(a) Find the distance of travel between t_0 and t_2.

(b) Find the average velocity between t_0 and t_2.

Example 2-8

Consider the graph of velocity vs. time in Fig. 2-6.

(a) Find the distance of travel between t_0 and t_2.

(b) Find the acceleration.

(c) Find the speed at $t = 6$ s.

(d) Describe the motion in words, commenting on the object's speed and its direction of motion.

Copyright © by June Chung – All rights reserved. Reproduction of this publication without the written permission of the copyright owner is strictly prohibited.

Key Point 2.8: Free-Falling Objects

motion of a free-falling object

An object dropped or thrown vertically downward always falls towards the ground because of gravity. Such an object, which is only under the influence of the gravitational force, is called a **free-falling object**. All free-falling objects experience a constant acceleration of 9.8 m/s^2 towards the center of the earth. Therefore, the speed of a free-falling object increases by 9.8 m/s every second. For example, an object thrown down with an initial speed of 2.0 m/s will be traveling at 11.8 m/s after one second, 21.6 m/s after two seconds, and so on.

motion of objects moving through the air

In fact, all objects traveling through the air experience a constant acceleration of 9.8 m/s^2 towards the center of the earth whether they are rising or falling. For instance, an object thrown vertically upwards at 12.6 m/s accelerates vertically downwards even while it is rising. To find the velocity of the object after one second, assume the upward direction as positive and apply Eq. (2.4):

$$\vec{a} = \frac{\vec{v}_f - \vec{v}_i}{\Delta t} = \frac{\vec{v}_f - (+12.6)}{1.0} = -9.8 \text{ m/s}^2$$

and we find $\vec{v}_f = 2.8$ m/s, upward. Similarly, you can find the object's speed is zero after 1.3 seconds, which implies that it becomes motionless at that exact instant. It is also at this instant that the object attains the highest elevation during its flight. The object then falls back towards the ground, its speed increasing by 9.8 m/s every second until reaching the ground.

> - Objects moving freely through the air always experience a constant acceleration of 9.8 m/s^2 towards the center of the earth.
> - Objects in mid-air lose speed while rising and gain speed while falling; they have the minimum speed at the highest elevation of the flight.

Copyright © by June Chung – All rights reserved. Reproduction of this publication without the written permission of the copyright owner is strictly prohibited.

Key Point 2.9: **Kinematic Equations**

When an object accelerates uniformly, its velocity vs. time graph is a straight line, and we can derive some useful equations to analyze such motions. Figure 2-7 shows an object uniformly accelerating in the positive direction (since the slope is positive).

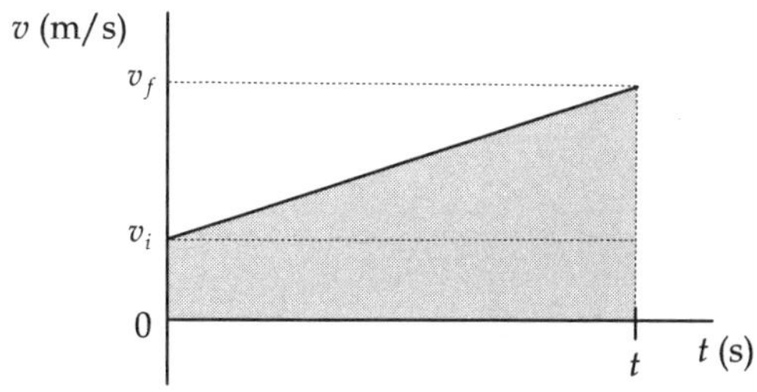

Figure 2-7: A graph of v vs. t for a uniformly accelerated motion

derivation of kinematic equations

The slope of a velocity vs. time graph represents the acceleration (see Key Point 2.7), which we can express as

$$\vec{a} = \frac{\text{rise}}{\text{run}} = \frac{\vec{v}_f - \vec{v}_i}{t - 0} = \frac{\vec{v}_f - \vec{v}_i}{t} \quad \ldots (2.5)$$

Solving Eq. (2.5) for $\vec{v}_f$, we obtain the following equation:

$$\vec{v}_f = \vec{v}_i + \vec{a}t \quad [\text{m/s}] \quad \ldots (2.6)$$

The displacement of an object is represented by the area under the graph. In Fig. 2-7, it is the sum of the areas of the rectangle and the triangle:

$$\vec{d} = \vec{v}_i t + \frac{1}{2}t\left(\vec{v}_f - \vec{v}_i\right) \quad \ldots (2.7)$$

Since $\vec{v}_f - \vec{v}_i = \vec{a}t$ from Eq. (2.6), Eq. (2.7) can be re-written as

$$\vec{d} = \vec{v}_i t + \frac{1}{2}\vec{a}t^2 \quad [\text{m}] \quad \ldots (2.8)$$

Now, you can combine Eqs. (2.6) and (2.8) to produce another equation:

$$v_f^2 = v_i^2 + 2\vec{a}\cdot\vec{d} \quad [\text{m}^2/\text{s}^2] \quad \ldots (2.9)$$

The derivation of Eq. (2.9) will be left as an exercise in Example 2-9.

condition on kinematic equations

All the equations above are valid for any object experiencing a constant acceleration (which may be positive, negative, or zero) and are useful in analyzing uniformly accelerated motion. Always make sure the acceleration is constant before you start using the equations. Usually

Copyright © by June Chung – All rights reserved. Reproduction of this publication without the written permission of the copyright owner is strictly prohibited.

you can use them for objects thrown into the air and objects moving on level ground.

> Is the acceleration of an object constant?
>
> ➢ If yes, you may use kinematic equations!

Example 2-9

Derive Eq. (2.9) by solving Eq. (2.6) for t first and then substituting the resulting expression into Eq. (2.8).

Copyright © by June Chung - All rights reserved. Reproduction of this publication without the written permission of the copyright owner is strictly prohibited.

Key Point 2.10: **How to Solve Kinematic Problems**

When solving kinematic problems, you probably will need to use the kinematic equations introduced in Key Point 2.9. Since there are a number of equations to choose from, it can be difficult to decide which equation(s) to use. Therefore, you need a systematic approach to solve these problems. Here is the solution procedure in four steps:

1. Read the problem statement carefully and make sure it is about uniformly accelerated motion. Draw a simple diagram that describes the motion, if one is not given.
2. Determine the positive direction of motion. It is not important which direction you set as positive, but it is crucial that you determine the values of displacement, velocity, and acceleration based on the chosen positive direction. For example, for an object moving in mid-air, its acceleration will be negative if you set the upward direction as positive, and will be positive if you set the downward direction as positive.
3. Choose an appropriate kinematic equation. The equation of choice depends on which variable you need to solve for and what data are given to you. For instance, if you need to solve for time of travel and you are given the final velocity, use Eq. (2.6). However, if you are given the displacement instead, use Eq. (2.8).

> - Need to solve for time?
>
>   Use $\vec{v}_f = \vec{v}_i + \vec{a}t$ or $\vec{d} = \vec{v}_i t + \frac{1}{2}\vec{a}t^2$
> - Need displacement?
>
>   Use $\vec{d} = \vec{v}_i t + \frac{1}{2}\vec{a}t^2$ or $v_f^2 = v_i^2 + 2\vec{a}\cdot\vec{d}$
> - Need final velocity?
>
>   Use $\vec{v}_f = \vec{v}_i + \vec{a}t$ or $v_f^2 = v_i^2 + 2\vec{a}\cdot\vec{d}$
>
> See what data is given to you and decide which equation to use!

Copyright © by June Chung – All rights reserved. Reproduction of this publication without the written permission of the copyright owner is strictly prohibited.

4. Substitute the data into the chosen equation and make sure you have only one unknown appearing in the equation. Then solve for the unknown. Think about whether your answer makes sense.

Again, watch out for the <u>vector</u> quantities such as displacement, velocity, and acceleration. You must think about their directions and attach the proper sign (that is, + or –) to their magnitudes. If you can do this, most kinematics problems should be fairly straightforward.

> ➢ Watch out for the directions of the vectors!

Example 2-10

An airplane, initially at rest, accelerates uniformly at 2.5 m/s^2 in order to take off at the end of the runway. If it needs a speed of at least 120 km/h to leave the ground, what is the minimum length of the runway? (Be careful of the units!)

Example 2-11

An object is dropped from the observation deck of the CN Tower, located 342 m above the ground. Ignoring air resistance, find (a) the displacement after 4.0 seconds; (b) the time it takes the object to travel half the distance to the ground; and (c) the average velocity for the entire flight.

Example 2-12

A ball, initially 1.2 m above the ground, is thrown vertically upward into the air with the speed of 15 m/s. Find (a) the maximum height the ball reaches; (b) the velocity after 2.2 seconds; and (c) the total time of flight.

Copyright © by June Chung – All rights reserved. Reproduction of this publication without the written permission of the copyright owner is strictly prohibited.

ANSWER KEY TO EXAMPLES

Example 2-1

	Initial position	Final position	Displacement
(a)	1.1 m above ground	Surface of ground	1.1 m, downward
(b)	0.5 m above ground	0.8 m above ground	0.3 m, upward

(c) The explorer eventually ends up 5 km south of where he or she started the journey. The displacement is thus 5 km, south.

Example 2-2

Tommy's initial and final positions are the same. His displacement is thus zero (no need for direction). His distance of travel is 3.2 km for the round-trip.

Example 2-3

(a) After a half revolution, Becky will be at the highest point of the Ferris wheel. Her final position will therefore be exactly two radii above the initial position. Her displacement is then 50 m, upward.

Meanwhile, her distance of travel is half the circumference of the wheel:

$$\text{distance of travel } = \pi(25) = 79 \text{ m}$$

(b) Since it takes 5.0 minutes to make one complete revolution, it will take her 2.5 minutes, or 150 seconds, to get to the highest point of the Ferris wheel. Her average velocity is therefore

$$\text{average velocity } = \frac{50}{150} = 0.33 \text{ m/s, upward}$$

Since Becky's distance of travel during the half revolution is 79 m, her average speed is

$$\text{average speed } = \frac{79}{150} = 0.52 \text{ m/s}$$

Example 2-4

(a) Cathy's distance of travel is 200 km. Thus, her average speed is

$$\text{average speed } = \frac{200}{2.50} = 80.0 \text{ km/h}$$

Cathy's displacement is 200 km, south. Therefore, her average velocity is 80.0 km/h, south.

(b) During the return trip, Cathy travels a total of 300 km in 4.00 hours. Her average speed is thus

Copyright © by June Chung – All rights reserved. Reproduction of this publication without the written permission of the copyright owner is strictly prohibited.

$$\text{average speed} = \frac{300}{4.00} = 75.0 \text{ km/h}$$

Cathy's going back to her friend's house to get her book does not affect her displacement, which is 200 km, north. Consequently, her average velocity is 50.0 km/h, north.

Example 2-5

First, set the positive direction as north. Then the initial velocity becomes +55 km/h, and the final velocity becomes -30 km/h. Converting the unit from km/h to m/s, we have

$$\vec{v}_i = +15 \text{ m/s, and } \vec{v}_f = -8.3 \text{ m/s}$$

According to Eq. (2.4), the average acceleration is

$$\vec{a} = \frac{\Delta\vec{v}}{\Delta t} = \frac{-8.3 - 15}{15} = -1.6 \text{ m/s}^2$$

The minus sign indicates that the direction is south. Therefore, the average acceleration is 1.6 m/s^2, south.

Note: You may set the positive direction as south instead of north. This changes the signs of the velocities:

$$\vec{v}_i = -15 \text{ m/s, and } \vec{v}_f = +8.3 \text{ m/s}$$

and the average acceleration becomes

$$\vec{a} = \frac{\Delta\vec{v}}{\Delta t} = \frac{+8.3 - (-15)}{15} = +1.6 \text{ m/ s}^2$$

This time the plus sign indicates that the direction is south. Therefore, we still get the same answer, which is 1.6 m/s^2, south.

Example 2-6

Set the positive direction as east. Then the initial velocity is +25 m/s, and the first acceleration +2.0 m/s^2. Using Eq. (2.4), we can find the object's velocity after the first acceleration:

$$\vec{a}_1 = +2.0 = \frac{\vec{v} - (+25)}{23}; \quad \vec{v} = +71 \text{ m/s}$$

Since the velocity is positive, the object is still moving east after the first acceleration. We can find the object's final velocity after the second acceleration similarly. Note that the second acceleration is -5.1 m/s^2 since its direction is west.

$$\vec{a}_2 = -5.1 = \frac{\vec{v}_f - (+71)}{11}; \quad \vec{v}_f = +15 \text{ m/s}$$

Hence, the final velocity of the object is 15 m/s, east.

Copyright © by June Chung – All rights reserved. Reproduction of this publication without the written permission of the copyright owner is strictly prohibited.

Example 2-7

(a) The distance of travel between t_0 and t_1 is 12 m, and that between t_1 and t_2 is 16 m. Thus, the total distance of travel is 28 m.

(b) The total displacement between t_0 and t_2 is 4.0 m in the negative direction. Divide it by the time of travel to obtain the average velocity:

$$\vec{v} = \frac{-4.0}{7.0} = -0.57 \text{ m/s}$$

or 0.57 m/s in the negative direction.

Example 2-8

(a) The distance of travel between t_0 and t_1 is 125 m, and that between t_1 and t_2 is 80 m. Thus, the total distance of travel is 205 m.

(b) The acceleration is the slope of the graph, which is

$$\vec{a} = \frac{-40-50}{9.0-0.0} = -10 \text{ m/s}^2$$

or 10 m/s^2 in the negative direction.

(c) The velocity decreases by 10 m/s every second from 50 m/s at $t = 0$. Hence, the velocity at $t = 6$ s is -10 m/s. The speed at $t = 6$ s is then the magnitude of the velocity, which is 10 m/s.

(d) The object starts out by moving in the positive direction at 50 m/s. During the first five seconds, it continues moving in the positive direction but constantly loses speed, and eventually it stops momentarily at $t = 5$ s. Then it reverses its direction of motion. It continues moving in the negative direction during the next four seconds while constantly gaining speed.

Example 2-9

Solve Eq. (2.6) for t:

$$t = \frac{\vec{v}_f - \vec{v}_i}{\vec{a}} \qquad \dots (1)$$

Substitute Eq. (1) into Eq. (2.8):

$$\vec{d} = \vec{v}_i\left(\frac{\vec{v}_f - \vec{v}_i}{\vec{a}}\right) + \frac{1}{2}\vec{a}\left(\frac{\vec{v}_f - \vec{v}_i}{\vec{a}}\right)^2 \qquad \dots (2)$$

Multiply both sides of Eq. (2) by $2\vec{a}$:

$$2\vec{a}\cdot\vec{d} = 2\vec{v}_i(\vec{v}_f - \vec{v}_i) + (\vec{v}_f - \vec{v}_i)^2 \qquad \dots (3)$$

Expand Eq. (3):

Copyright © by June Chung – All rights reserved. Reproduction of this publication without the written permission of the copyright owner is strictly prohibited.

$$2\vec{a}\cdot\vec{d} = 2\vec{v}_i\cdot\vec{v}_f - 2v_i^2 + v_f^2 - 2\vec{v}_i\cdot\vec{v}_f + v_i^2 \qquad \ldots (4)$$

After simplifying Eq. (4), we obtain

$$2\vec{a}\cdot\vec{d} = v_f^2 - v_i^2 \qquad \ldots (5)$$

which can be rearranged to become Eq. (2.9).

Example 2-10

The airplane is accelerating uniformly, so we can use the kinematic equations. Below is the diagram depicting the motion:

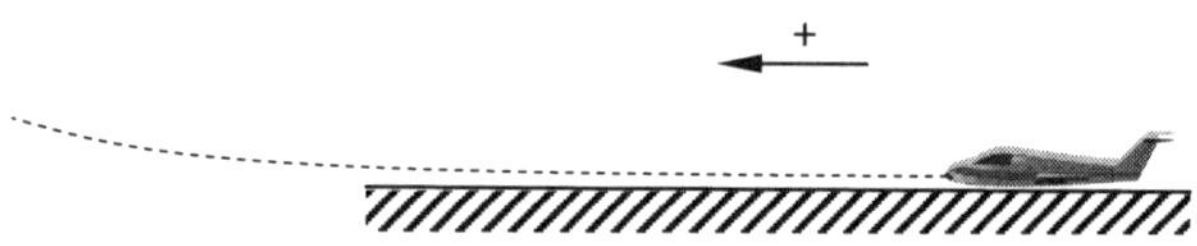

If we set the positive direction as shown in the diagram, the initial velocity, the acceleration, and the final velocity can be expressed as

$$\vec{v}_i = 0 \text{ m/s};\ \vec{a} = +2.5 \text{ m/s}^2;\ \vec{v}_f = +120 \text{ km/h} = +33 \text{ m/s}$$

Since we are solving for the displacement and given the final velocity, choose Eq. (2.9) and substitute the values into the equation:

$$33^2 = 0^2 + 2(2.5)\cdot\vec{d}$$

Solving for the displacement, we obtain $\vec{d} = +220$ m. Thus, the minimum length of the runway should be 220 meters.

Example 2-11

Because the object is dropped, its initial velocity is zero. If we set the downward direction as positive, the acceleration of the object is +9.8 m/s^2.

(a) To find the displacement after 4.0 seconds, use Eq. (2.8):

$$\vec{d} = 0(4.0) + \frac{1}{2}(+9.8)(4.0)^2 = 78 \text{ m, downward}$$

(b) When the object is halfway to the ground, its displacement is +171 m. Again, use Eq. (2.8), but solve for t this time:

$$+171 = 0\cdot t + \frac{1}{2}(+9.8)t^2;\quad t = 5.9 \text{ s}$$

(c) Knowing the displacement, which is +342 m, we only need to find the time of travel for the entire flight. As we did in (b), use Eq. (2.8):

$$+342 = 0\cdot t + \frac{1}{2}(+9.8)t^2;\quad t = 8.4 \text{ s}$$

Therefore, the average velocity of the object is

$$\vec{v}_{avg} = \frac{\vec{d}}{t} = \frac{+342}{8.4} = 41 \text{ m/s, downward}$$

Copyright © by June Chung – All rights reserved. Reproduction of this publication without the written permission of the copyright owner is strictly prohibited.

Example 2-12

Since all objects traveling in mid-air experience a constant acceleration of 9.8 m/s² towards the ground, we can use the kinematic equations. If we set the upward direction as positive, the initial velocity and the acceleration can be written as

$$\vec{v}_i = +15 \text{ m/s}; \ \vec{a} = -9.8 \text{ m/s}^2$$

(a) At the maximum height, the ball is motionless, meaning its velocity is zero. Thus, $\vec{v}_f = 0$ m/s. Use Eq. (2.9) to find the ball's displacement:

$$0^2 = 15^2 + 2(-9.8) \cdot \vec{d}$$

The displacement of the ball is +11.5 m, or 11.5 m upwards. Thus, the maximum height of the ball is 11.5 m above the initial position, or 12.7 m above the ground.

(b) Because we need to solve for the final velocity and are given time, use Eq. (2.6):

$$\vec{v}_f = +15 + (-9.8) \cdot (2.2)$$

The final velocity is -6.6 m/s, or 6.6 m/s, downward.

(c) When the ball hits the ground, its displacement is 1.2 m, downward, or -1.2 m. Since we know the displacement and need the time of flight, use Eq. (2.8):

$$-1.2 = +15t + \frac{1}{2}(-9.8) \cdot t^2$$

Solve for t using the quadratic formula, and we obtain

$$t = -0.08 \text{ s}, 3.1 \text{ s}$$

Since we cannot have a negative time of flight, the total time of flight is 3.1 s.

Copyright © by June Chung – All rights reserved. Reproduction of this publication without the written permission of the copyright owner is strictly prohibited.

PROBLEMS ON CHAPTER 2: KINEMATICS

(Answer Key on p. 223)

1 - 10. Fill in the blanks

1. A _________ vector refers to an object's location.
2. The _________ of an object is a change in its position.
3. The _________ of travel is the length of an actual path taken by an object.
4. The _________ of an object indicates the rate at which it is changing its position.
5. The _________ of an object is a scalar that indicates how fast it is traveling.
6. An _________ vector indicates the rate at which an object is changing its velocity.
7. The area under a velocity vs. time curve represents _________.
8. The slope of a position vs. time curve represents _________.
9. The slope of a velocity vs. time curve represents _________.
10. An object that experiences only the force of gravity is a _________ object.

11 - 20. True/False questions

11. _____ The position of an object depends on the location of the observer.
12. _____ To find the displacement of a body, we only need to know its initial and final positions.
13. _____ If the displacement of an object is zero, it means the object has not moved.
14. _____ When the displacement of a body is zero, its average velocity is also zero.
15. _____ The speed of an object may vary even though its velocity does not change.
16. _____ The direction of an object's acceleration is always the same as that of its velocity.
17. _____ The area under a position vs. time graph represents displacement.
18. _____ A free-falling body always experiences a constant acceleration of 9.8 m/s^2.
19. _____ The kinematic equations can be applied to analyze the motion of objects accelerating uniformly.
20. _____ When two objects with different masses are dropped simultaneously from the same height, the heavier object hits the ground first.

Copyright © by June Chung – All rights reserved. Reproduction of this publication without the written permission of the copyright owner is strictly prohibited.

21 - 28. Multiple choice questions

21. An object is moving at constant speed along a curved path. Which of the following statements is true?

    a) Its velocity remains constant.
    b) Its velocity constantly changes.
    c) Its velocity changes once in the middle of the curve.
    d) We cannot determine whether its velocity changes unless the exact path of travel is known.

22. An object, initially at rest, moves for 3 seconds and settles at an unknown location. Which of the following statements is true?

    a) The displacement is zero.
    b) The distance of travel is zero.
    c) The displacement is not zero.
    d) The distance of travel is not zero.

23. Which of the following is true for an object accelerating at 25 m/s^2, east?

    a) The direction of its displacement is east.
    b) The direction of its velocity is east.
    c) Its speed is increasing.
    d) Its instantaneous velocity is changing.

24. Which of the following is true for an object thrown straight up in the air?

    a) The magnitude of its acceleration is 9.8 m/s^2 and the direction is up.
    b) The magnitude of its acceleration is 9.8 m/s^2 and the direction is down.
    c) There is no acceleration.
    d) The acceleration cannot be determined unless we know its initial speed.

25. An object is thrown vertically upward at 11 m/s from the ground and falls back to the ground a short time later. Which of the following statements is false?

    a) The displacement of the object is zero.
    b) The final speed of the object at the ground is equal to its initial speed.
    c) The final velocity of the object at the ground is equal to its initial velocity.
    d) The final acceleration of the object at the ground is equal to its initial acceleration.

Copyright © by June Chung - All rights reserved. Reproduction of this publication without the written permission of the copyright owner is strictly prohibited.

26. Which of the following graphs shows a motion with non-zero uniform acceleration?

a) 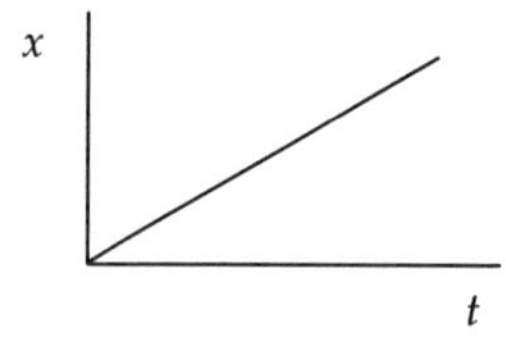

b) 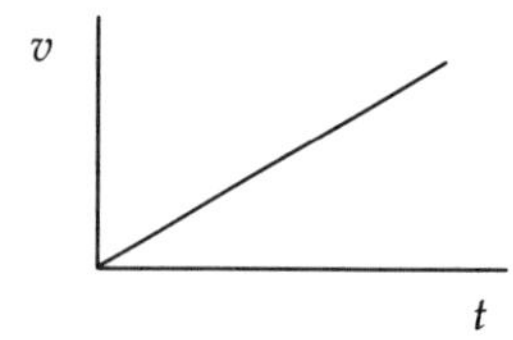

c) 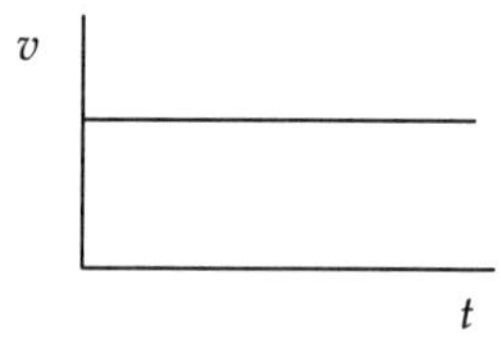

d) 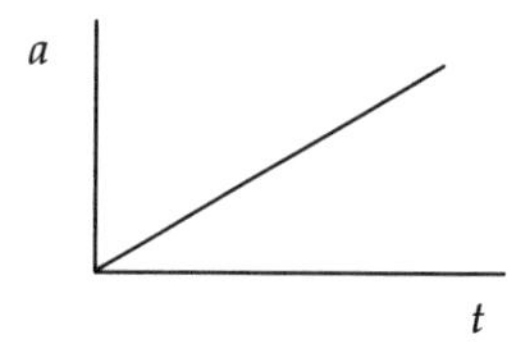

27. Which of the following position vs. time graphs represents the motion of an object thrown vertically upward?

a) 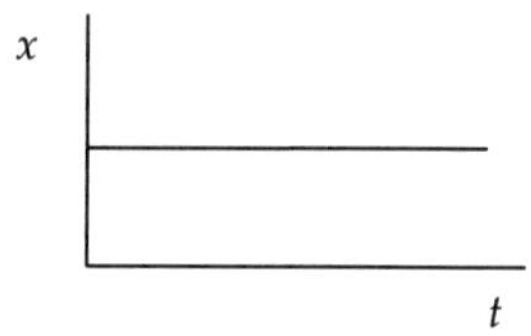

b) 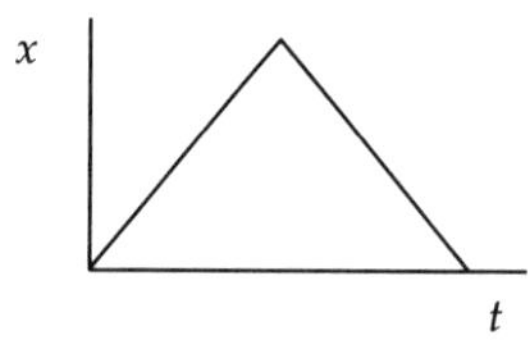

c) 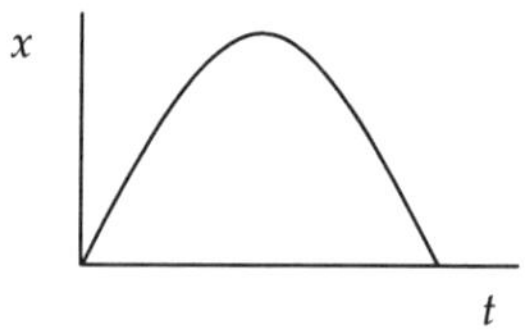

d) 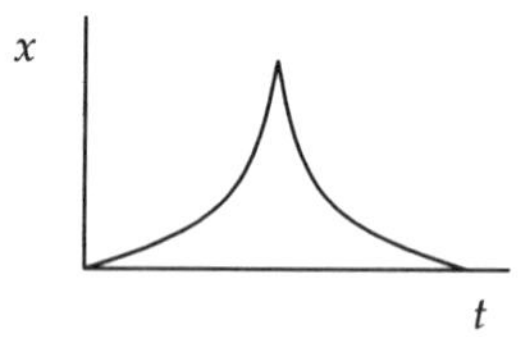

28. Which of the following describes the motion shown in the position vs. time graph below?

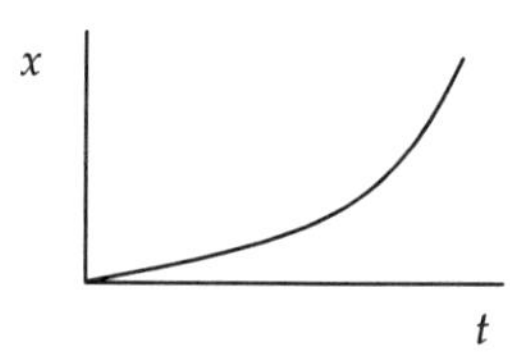

a) The displacement is constant.
b) The velocity is constant.
c) The velocity is changing.
d) The acceleration is zero.

Copyright © by June Chung – All rights reserved. Reproduction of this publication without the written permission of the copyright owner is strictly prohibited.

29 – 40. Written-response questions

29. Your physics teacher ran a mini-marathon where she had to run from point A to point B and return to point A. From point A to B, she ran at a constant speed of 2.5 m/s. On the way back to point A, she ran at 2.0 m/s. What was her average speed for the entire course?

30. The velocity vs. time graphs of two objects in one-dimensional motion are shown below.

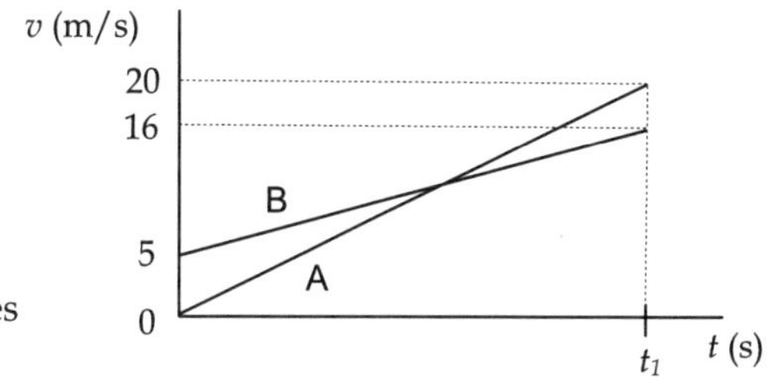

a) Compare the displacements of the objects between $t = 0$ and $t = t_1$.

b) Compare the average velocities of the objects between $t = 0$ and $t = t_1$.

c) Compare the instantaneous velocities of the objects at $t = t_1$.

d) Compare the average accelerations of the objects between $t = 0$ and $t = t_1$.

31. A motorcycle, initially at rest, accelerates at 2.50 m/s^2 until it reaches the speed limit of 60 km/h. Then the motorcycle travels for 2.0 minutes at that speed. Find the total distance of travel.

32. Plot the velocity vs. time graph of the motorcycle in Problem 31. Indicate the displacement and the acceleration of the motorcycle in the plot.

33. An object initially moving to the right at 15 m/s accelerates uniformly for 5.0 seconds, after which the object is traveling at 35 m/s to the left.

a) What was the average acceleration of the object?

b) How long did it take for the object to come to rest momentarily?

c) What was the total displacement of the object during the 5.0 second period? What about its distance of travel?

Copyright © by June Chung - All rights reserved. Reproduction of this publication without the written permission of the copyright owner is strictly prohibited.

34. An object accelerates uniformly from rest at 3.0 m/s^2 for 2.4 seconds. If a second object is to cover the same distance within 1.2 seconds with the initial velocity of 5.3 m/s, what should be its minimum acceleration?

35. A bullet traveling at 315 m/s strikes a wooden block at rest. Upon impact, the bullet is embedded within the block, and they start to slide on a horizontal surface at 17 m/s. However, they soon come to a full stop after traveling 22 m because of friction. Determine the deceleration of the block. (Assume the block does not move until the bullet is fully embedded and comes to rest within it.)

36. You throw a 3.5-kg object straight up into the air. Sketch qualitatively the speed vs. time curve for the object while it is in the air, and indicate the point on the curve at which the ball reaches the maximum height.

37. A ball is tossed vertically upward at 21 m/s. Find the distance the ball travels <u>during the third second</u> of its flight.

38. A balloon is floating 15 m above the ground. If you want to hit the balloon with a baseball thrown from the ground, calculate the minimum initial speed of the ball. (Assume you throw the ball vertically.)

39. A stone is thrown vertically upward at 12 m/s from the top of a building. If it hits the ground after 5.1 seconds, how tall is the building? What is the average speed and the average velocity of the stone?

40. An object is dropped from the top of a building. As it falls, it passes a 2.4-m tall window in 0.15 seconds. Determine the height from which the object was dropped relative to the top of the window.

Copyright © by June Chung – All rights reserved. Reproduction of this publication without the written permission of the copyright owner is strictly prohibited.

CHAPTER 3

DYNAMICS

Now that we understand about motion, we can turn our attention to Dynamics, which explains why objects move. Again your experiences have taught you a lot about this topic already. For example, if you want to move a ball, you need to force it to move by kicking it or hitting it with a bat. If you push the ball in the wrong direction, it will go in that direction. If you hit it harder, it will go farther and faster, but eventually it will stop moving. Also, if you kick a car or a piano, it will not move, and indeed your foot may suffer an injury. These well understood ideas form the basis of Newton's laws. They enable us to explain when motion occurs, why some objects accelerate, and how objects interact with each other.

In this chapter, you will study:

- ✓ force
- ✓ Newton's three laws of motion
- ✓ various kinds of force: gravitational, normal, tension, and friction force
- ✓ how to draw a free-body diagram
- ✓ how to solve dynamic problems

Copyright © by June Chung – All rights reserved. Reproduction of this publication without the written permission of the copyright owner is strictly prohibited.

Key Point 3.1: **Force ($\vec{F}$)**

force

Force is a push or a pull, and the agent of the force may be visible (for example, a ball hitting a window) or invisible (for example, gravity pulling the ball down). It is a vector quantity, and its SI unit is the Newton [N]. If an object is acted upon by a single force, it always accelerates in the direction of that force. For instance, objects moving freely in mid-air experience only the gravitational force, which pulls them downward. Accordingly, they always accelerate downward as long as they are in the air.

effect of a single force on motion of an object

effect of balanced forces on motion of an object

However, if an object is acted upon by two or more forces, it may not accelerate at all. For instance, a book sitting on a level surface experiences two forces: the force of gravity pulling it down, and the force exerted by the surface pushing the book up. In this case, the two forces cancel each other out, and we say that the two forces are balanced. Accordingly, the book maintains its state of motion, which is being at rest.

effect of unbalanced forces on motion of an object

On the other hand, if two forces acting on an object are not balanced, the object always accelerates in the direction of the force with the larger magnitude. For example, when you pick up a volleyball lying on the ground, the ball feels two forces: the force of gravity pulling it down, and the upward force that you apply. This time, the magnitude of the applied force is greater than that of the gravitational force. Consequently, the ball accelerates in the direction of the applied force, or upward. The greater the difference in magnitude between the two forces, the greater the acceleration of the ball.

In summary, to determine whether an object accelerates or not, you have to determine whether the forces acting on the object are balanced. We will study the relationship between force and motion using Newton's laws of motion in the next three Key Points.

Copyright © by June Chung – All rights reserved. Reproduction of this publication without the written permission of the copyright owner is strictly prohibited.

- Force is a push or a pull, which can accelerate a body.
- If only a single force acts on a body, it accelerates in the direction of that force.
- If two or more forces act on a body, determine whether the forces are balanced.

  ⇨ If balanced, the body does not accelerate.
  If not balanced, the body accelerates.

Example 3-1

Determine whether each of the following objects will accelerate.

(a) A rock that is dropped from the top of a tall building.

(b) An automobile parked on level ground.

(c) A baseball being thrown by a pitcher.

(d) An object acted on by two forces: 20 N, north, and 20 N, south.

(e) An object acted on by two forces: 20 N, east, and 10 N, west.

(f) An object acted on by three forces: 15 N, north, 15 N, south, and 15 N, west.

Copyright © by June Chung – All rights reserved. Reproduction of this publication without the written permission of the copyright owner is strictly prohibited.

Key Point 3.2: Newton's First Law of Motion

When the forces on an object are balanced, the object does not accelerate. Its acceleration being zero, the object will travel with constant velocity or stay at rest. If it is moving, it will travel at constant speed in a straight line. This is summarized in **Newton's first law of motion**: if the forces acting on a body are balanced, the body maintains its state of motion.

Newton's First Law

Note that the converse is also true: if an object is moving with constant velocity or is motionless, the forces acting on it are balanced. Thus, from now on, whenever you observe an object moving with constant velocity or at rest, you should conclude that its acceleration is zero and the forces on it are balanced.

> ➢ Newton's first law of motion:
> "When forces acting on an object are balanced, the object maintains its state of motion."
>
> ➢ These three statements have the same meaning:
> 1. An object is moving with constant velocity.
> 2. An object is not accelerating.
> 3. Forces on an object are balanced.

According to Newton's first law, a moving object will continue its motion along a straight line with constant speed as long as the forces on it are balanced. This tendency of a body to maintain its current state of motion is called **inertia**. The greater the inertia of an object, the harder it is to change its state of motion. In other words, an object with a large inertia is 'stubborn.' When it is moving, it wants to keep moving, and when it is not moving, it wants to stay at rest. Since the concept of inertia is closely tied to Newton's first law, this law is sometimes called the law of inertia.

inertia

example situations involving inertia

Some of the situations that can be explained by the concept of inertia are as follows:

1. When you are aboard a bus which suddenly stops, you can feel your body lean forward (because your body wants to maintain its forward motion).

Copyright © by June Chung – All rights reserved. Reproduction of this publication without the written permission of the copyright owner is strictly prohibited.

2. When the elevator starts to accelerate downward, your body feels lighter (because your body wants to stay at rest).
3. When a dog is wet, it shakes its fur in one direction to get the water molecules moving in that direction, and then when the dog shakes in the opposite direction, the water molecules fly off the dog's body (because they want to continue moving in the initial direction).

> - Inertia is the tendency of a body to maintain its current state of motion.
> - The greater the inertia of a body, the harder it is to change its velocity.

Example 3-2

Helen pushes an eraser on a flat surface so that it moves along a straight line at constant speed. Two of the forces acting on the eraser are the force of gravity acting downward and Helen's applied force acting in the direction of motion. Identify additional forces on the eraser, and indicate which forces balance the force of gravity and the applied force, respectively.

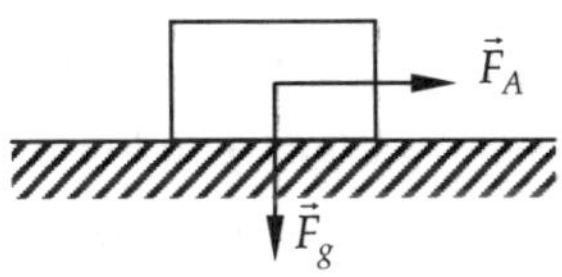

Copyright © by June Chung – All rights reserved. Reproduction of this publication without the written permission of the copyright owner is strictly prohibited.

Key Point 3.3: Newton's Second Law of Motion

Newton's first law of motion explains why an object maintains its state of motion. It does so because the forces acting on it are balanced. The second law then answers the question, "What happens to an object's motion if these forces are not balanced?" Now the object will accelerate, and **Newton's second law** states that the acceleration is obtained according to

Newton's second law

$$\vec{a} = \frac{\vec{F}_{net}}{m} \quad [\text{N/kg} = \text{m/s}^2] \qquad \ldots (3.1)$$

$\vec{F}_{net}$ is the vector sum of all the individual forces acting on an object and is called the '**net force**.' According to Eq. (3.1), a 5.0-N net force acting on a 1.0-kg body will cause it to accelerate at 5.0 m/s^2. Thus, we can define 1 N as the amount of force required to accelerate a 1-kg mass by 1 m/s^2. (Hence 1 N is equivalent to 1 kg·m/s^2.)

acceleration is in the same direction as net force

Equation (3.1) shows that the direction of the acceleration is always the same as that of net force. Thus, if you can determine the direction of the net force acting on an object, you know the direction in which the object accelerates, and vice versa. For example, if you lift an object that weighs 250 N with a 310-N force, the net force on it is 60 N, up, and the object accelerates upward. Also, if you observe an object slowing down on a horizontal surface, you know that it is accelerating in the opposite direction to the motion, and, therefore, the net force on the object opposes the motion as well.

> ➢ Newton's second law of motion:
> When the forces on an object are not balanced, the object accelerates according to
>
> $$\vec{a} = \frac{\vec{F}_{net}}{m} \quad (\text{or } \vec{F}_{net} = m\vec{a})$$

Example 3-3

A 5.2-kg object experiences the forces shown below. Calculate its acceleration.

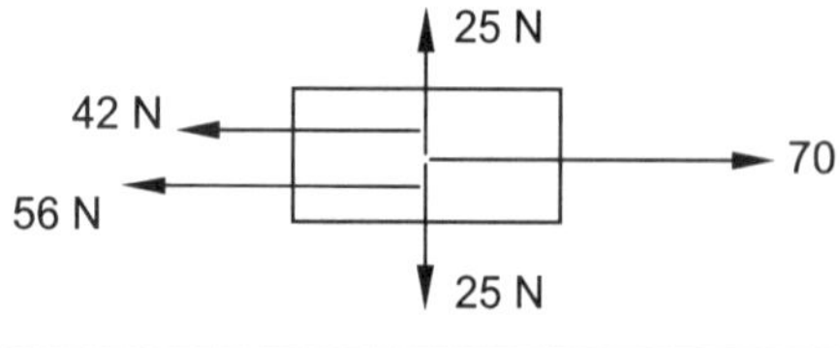

Copyright © by June Chung – All rights reserved. Reproduction of this publication without the written permission of the copyright owner is strictly prohibited.

Key Point 3.4: Newton's Third Law of Motion

Newton's third law

Newton's third law of motion explains the interaction between two objects when one exerts a force on the other. It states that, for every action, there is an equal and opposite reaction. For instance, if you push a wall with 100 N of force, the wall will push back on you with the same magnitude of force. Consequently, the harder you push the wall, the harder the wall pushes back. Likewise, the harder you kick a wall, the more your foot hurts! This law can be expressed in a mathematical equation as

$$\vec{F}_{on\,A\,by\,B} = -\vec{F}_{on\,B\,by\,A} \quad \ldots (3.2)$$

That is, a force on *A* applied by *B* is always accompanied by a reaction force on *B* applied by *A* with the same magnitude but in the opposite direction, as shown in Fig. 3-1. The force on *A*, not the force exerted by *A*, must be used when determining the acceleration of *A*.

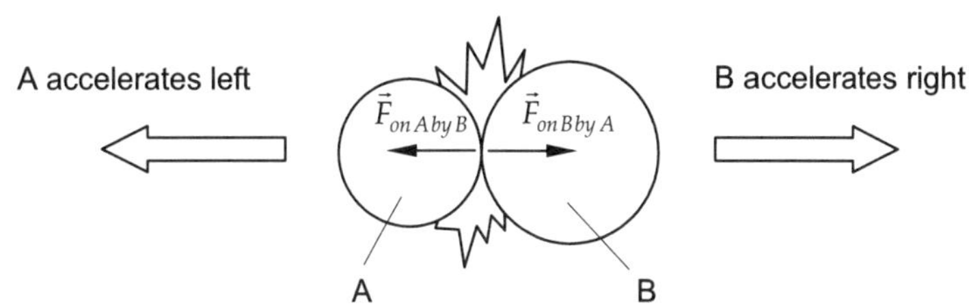

Figure 3-1: Action and reaction forces

example situations involving action and reaction

Situations involving the propulsion of an object can be explained by Newton's third law. Examples are:

1. Propulsion of a rocket: A rocket pushes gas back out through the nozzle, and the gas in turn pushes the rocket forward.
2. Propulsion of a boat: A boat propeller pushes water backwards, and the water in turn pushes the boat forward.
3. Walking: You push the ground back with your foot, and the ground in turn pushes you forward.

> ➢ Newton's third law of motion:
> "For every action, there is an equal and opposite reaction"
> ➢ Most situations involving the propulsion of an object can be explained by Newton's third law of motion.

Copyright © by June Chung - All rights reserved. Reproduction of this publication without the written permission of the copyright owner is strictly prohibited.

Example 3-4

Julie and Jack are initially skating at constant velocity when Jack suddenly pushes Julie from behind with a 130-N force. If Julie and Jack are 45 kg and 60 kg, respectively, find their accelerations while the force is being applied. (Hint: Use Newton's second and third laws.)

Copyright © by June Chung - All rights reserved. Reproduction of this publication without the written permission of the copyright owner is strictly prohibited.

Key Point 3.5: Gravitational Force ($\vec{F}_g$)

gravitational force

One of the fundamental forces in nature, **gravitational forces** act between all masses and are attractive. That is, every object exerts a gravitational force on every other object, which tends to draw them together. For instance, this book is pulling your body towards itself. So is the chair you are sitting on. However, you may usually ignore these gravitational forces because they are very small in magnitude unless a massive object or objects are involved. For example, the gravitational force that the earth exerts on your body is not negligible because of the earth's enormous mass. (Earth has about 10^{23} times the mass of an average adult.)

magnitude and direction of weight

In general, the gravitational force of the earth is the only gravitational force we need to be concerned about, and we call this force acting on a body its **weight** (on Earth). The magnitude of weight is obtained by multiplying the body's mass by the gravitational field strength of the earth, g, of 9.8 N/kg (or m/s^2). Thus, a 10-kg object, for example, weighs 98 N on the surface of the earth. The weight is always directed toward the ground, or the center of the earth. Note that weight is a force, and the correct unit for weight is thus the Newton [N], not the kilogram [kg]. (The kilogram is the standard unit for mass, which is a scalar quantity that refers to the amount of material in a body.)

> - Gravitational forces act between all masses and are attractive.
> - Weight is the force of gravity acting on a mass.
> $\vec{F}_g = m\vec{g}$ (g = 9.8 m/s^2 on the surface of the earth)
> - Mass refers to the amount of material in a body.

Note: The moon has the gravitational field strength of 1.6 N/kg. Thus, a 10-kg object weighs only 16 N on the surface of the moon.

Copyright © by June Chung – All rights reserved. Reproduction of this publication without the written permission of the copyright owner is strictly prohibited.

Key Point 3.6: Normal Force ($\vec{F}_N$)

normal force

Normal force is the force that a surface exerts on an object lying upon it. In order for the normal force to exist, an object must be in contact with the surface. As soon as an object loses contact with the surface, the normal force disappears. The direction of a normal force is always perpendicular to and away from the surface, as shown in Fig. 3-2.

direction of normal force

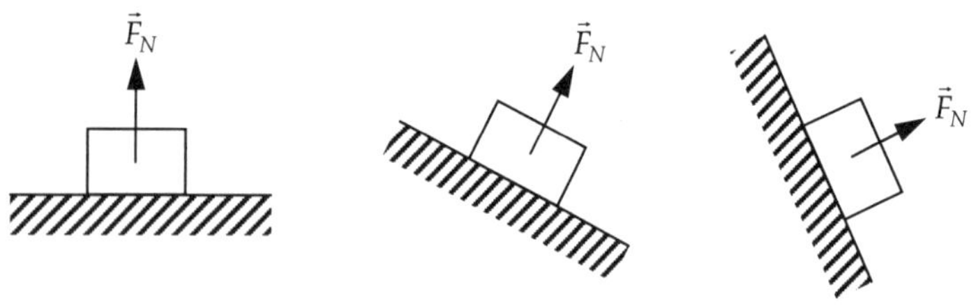

Figure 3-2: Directions of normal forces

magnitude of normal force in various situations

When an object sits on top of a level surface, it experiences a normal force, $\vec{F}_N$, whose magnitude is equal to that of its weight, $\vec{F}_g$ (Fig. 3-3 (a)). Now, if someone presses down on the object from the top, the magnitude of the normal force increases by the magnitude of the applied force, $\vec{F}_a$ (Fig. 3-3 (b)). If, on the other hand, someone tries to lift the object, the magnitude of the normal force decreases (Fig. 3-3 (c)) until the magnitude of the lifting force becomes equal to that of the weight. The moment the magnitude of the lifting force becomes greater than that of the weight, the object begins to accelerate upwards, losing contact with the surface. From that point on, the normal force on the object ceases to exist (Fig. 3-3 (d)).

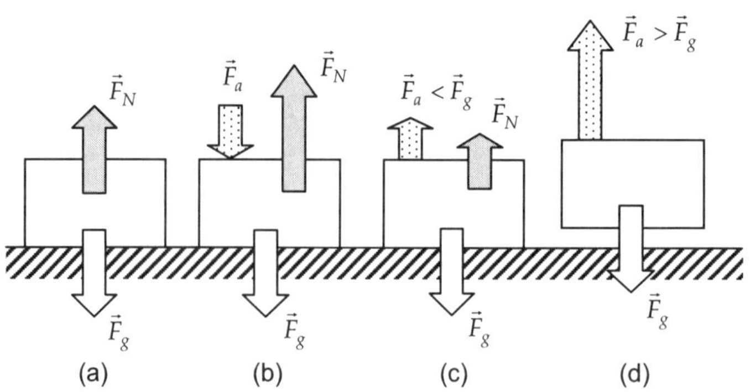

Figure 3-3: Normal forces in various situations

Copyright © by June Chung – All rights reserved. Reproduction of this publication without the written permission of the copyright owner is strictly prohibited.

The previous explanation can be used to prove that the normal force is not the reaction force to the force of gravity. If it were, then its magnitude should always be the same as that of the gravitational force, but we have seen that it can increase, decrease, or even disappear! For this reason, the normal force is sometimes called an accommodating force because its magnitude varies depending on the conditions.

> - Normal force is the force that a surface exerts on a body lying on it.
> - The normal force always acts perpendicular to the surfaces in contact and points away from those surfaces.

Example 3-5

An 18-kg dumbbell is lying on level ground. Find:

(a) its weight and the normal force.

(b) the normal force when you apply a 125-N lifting force on the dumbbell.

(c) the normal force when you increase the lifting force to 225 N.

Copyright © by June Chung – All rights reserved. Reproduction of this publication without the written permission of the copyright owner is strictly prohibited.

Key Point 3.7: Tension Force ($\vec{F}_T$)

tension force and its direction

Tension force is the force transmitted through a string when it is pulled tight. When you pull an object with a string, the direction of the tension force on an object is always along the string and away from the object.

magnitude of tension force

When a string is pulled tight, the tension force is the same along the total length of the string. Consequently, if two objects are connected by a string and are pulling on that string, they will experience exactly the same magnitude of tension force. However, if two different strings are connected to the same object, the tension forces in the strings are generally not equal to each other (as you will see in Example 3-6).

motion of two objects connected by a string

One important assumption in all problems involving strings is that the strings are not elastic. That is, they do not stretch. This implies that two masses connected by a string always travel the same distance, possess the same speed, and experience the same magnitude of acceleration while in motion. Hence, when solving a dynamic problem involving two masses connected by a string, you should set the magnitude of the acceleration of each mass equal to each other.

- Tension force is the force transmitted through a string when it is pulled tight.
- A tension force always acts along the string and points away from the object.
- When two masses are connected by a string, they move with the same speed and the same magnitude of acceleration.

Example 3-6

Two masses connected by a string are hanging from a ceiling by another string as shown. Given that m_1 = 15 kg and m_2 = 22 kg, find the tension force in each string.
(Hint: Use Newton's first law of motion.)

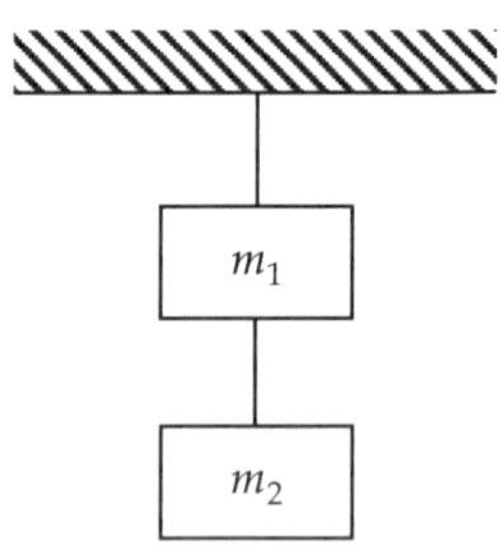

Copyright © by June Chung – All rights reserved. Reproduction of this publication without the written permission of the copyright owner is strictly prohibited.

Key Point 3.8: Friction Force ($\vec{F}_f$)

friction force

Friction force is the force that a surface exerts on an object as the object slides or tries to slide over it. While the normal force acts perpendicular to the surface, the friction force always acts parallel to the surface and to oppose the motion of the object. Therefore, the friction force vector is always perpendicular to the normal force vector, if both exist.

direction of friction force

magnitude of friction force

The magnitude of a friction force is the product of the coefficient of friction, μ, and the magnitude of the normal force:

$$F_f = \mu F_N \qquad \text{... (3.3)}$$

The **coefficient of friction** is determined by the nature of the materials in contact with each other, and the magnitude of the normal force indicates how closely those materials are in contact. Accordingly, the greater the normal force, the harder it is to slide one material over the other. Note that the friction force is independent of the surface areas in contact and independent of the speed of the motion.

determining friction force

For a moving object, apply Eq. (3.3) to calculate the magnitude of the friction force. The direction of the friction force is always opposite to that of the motion. However, when an object is at rest, you do not need any equation to find the friction force; simply apply Newton's first law to the stationary object.

- Friction force is the force that a surface exerts on a body as the body slides or tries to slide over it.
- The direction of a friction force is always parallel to the surface and opposes the motion of the body.

When a body is in motion...

- use $F_f = \mu F_n$ to find the friction force.

When a body is at rest (in the presence of friction)...

- Find the friction force by setting net force to zero.

Copyright © by June Chung – All rights reserved. Reproduction of this publication without the written permission of the copyright owner is strictly prohibited.

Example 3-7

A 3.5-kg object is at rest on an inclined surface as shown. There are three forces acting on the object: the force of gravity, the normal force, and the friction force. Find the magnitude of the friction force using Newton's first law. (Hint: The three force vectors, when added using the tip-to-tail method, form a right triangle.)

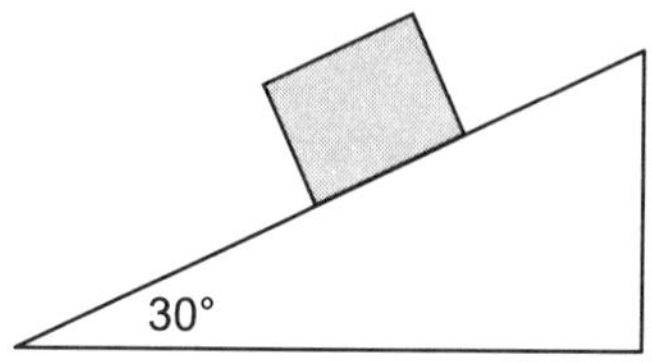

Example 3-8

Kathy is pulling on a rope connected to a 32-kg recycling bin full of empty bottles. The bin slides over the surface at constant speed. If the tension on the rope is 56 N, what is the coefficient of friction between the wheels and the surface? (Hint: Apply Newton's first law of motion.)

Copyright © by June Chung – All rights reserved. Reproduction of this publication without the written permission of the copyright owner is strictly prohibited.

Key Point 3.9: **Free-Body Diagram**

A **free-body diagram** shows an object of interest and all the forces acting on it (but not the ones exerted by it). You need to be able to draw free-body diagrams in various situations in order to understand how each force acting on a body contributes to its motion. When drawing a free-body diagram, each force should, if possible, point in the proper direction. You may assume all forces act through the object's center of mass by treating it as a single point.

algorithm for drawing a free-body diagram

Use the following procedure to draw a free-body diagram:

1. Draw the object of interest.
2. Draw the force of gravity pointing towards the center of the earth.
3. Look for things that the object is in contact with.
   - If the object is in contact with a surface, draw the normal force and the friction force as instructed in Key Point 3.6 and Key Point 3.8. Friction does not exist if the surface is smooth.
   - If the object is in contact with a string, draw the tension force as instructed in Key Point 3.7.
   - If the object is in contact with another object, draw the force applied by that object on the object of interest. Usually, the direction will be obvious from the problem statement.

The net force on the object is the sum of all the forces that appear in the free-body diagram. The net force vector itself should never appear in the diagram.

When drawing a free-body diagram...

- Always include the force of gravity.
- Include the normal force and the friction force if the object is in contact with a surface.
- Include the tension force if the object is in contact with a string.
- Include the applied force if the object is in contact with another body.
- Remember to include only the forces acting on the object, not exerted by the object!

Copyright © by June Chung – All rights reserved. Reproduction of this publication without the written permission of the copyright owner is strictly prohibited.

Example 3-9

Draw the free-body diagram of each object.

(a) A block is pulled up an inclined surface by a rope.

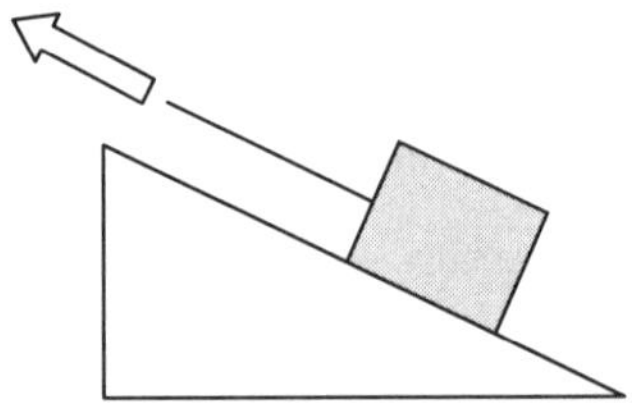

(b) Two masses connected by a string are pulled to the right at constant speed while experiencing friction.

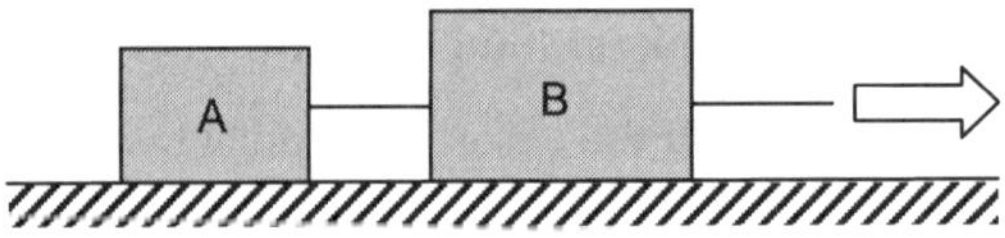

(c) A pendulum bob is held steady by a horizontal applied force.

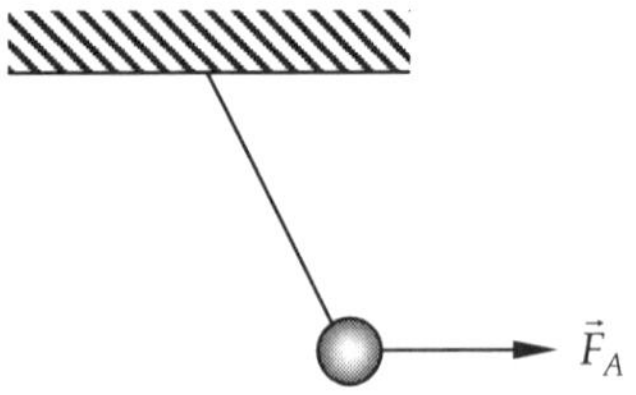

Copyright © by June Chung – All rights reserved. Reproduction of this publication without the written permission of the copyright owner is strictly prohibited.

Key Point 3.10: **How to Solve Dynamic Problems**

Before attempting to solve any dynamic problem, make sure you can draw free-body diagrams of objects in various situations. Constructing the correct free-body diagram is the most critical part of solving the problem, and the rest of the procedure is fairly straightforward. Keeping that in mind, follow the four steps below:

1. Read the problem statement carefully. Look for phrases such as "constant velocity" or "smooth surface." As soon as you see "constant velocity," set the acceleration of the object to zero and the net force on the object to zero. When you see "smooth surface," you know you don't have to worry about friction.

2. Draw a free-body diagram for each mass as instructed in Key Point 3.9. Even if you think you can solve the problem without it, don't take the chance! I cannot overemphasize this point. Get in the habit of always drawing a free-body diagram before applying Newton's laws.

3. Set the direction in which the object is moving as positive, and set up an equation of motion for each mass using Newton's second law:

$$\vec{F}_{net} = m\vec{a} \qquad \dots (3.4)$$

Use free-body diagram! → dynamic side | kinematic side ← Use kinematic equations!

To find the net force, go to the free-body diagram of the object. Add the forces opposing the motion and subtract it from the forces encouraging the motion. Forces perpendicular to the direction of motion do not enter into Eq. (3.4) because they must be balanced.

4. Check that there is only one unknown in the equation, and solve for it. (If there are two masses connected by a string, there should be two equations.) After you obtain an answer, think about whether it is reasonable. Remember that the direction of the acceleration is independent of the direction of motion. If the acceleration is in the positive direction, the object must gain speed. If it is in the negative direction, the object must slow down.

Copyright © by June Chung – All rights reserved. Reproduction of this publication without the written permission of the copyright owner is strictly prohibited.

Example 3-10

Kathy is pulling on a rope connected to a 32-kg recycling bin full of empty bottles. The coefficient of friction between the wheels and the surface is 0.23. If she exerts constant 75-N force on the rope, what is the acceleration of the bin?

Example 3-11

A system of masses shown below is pulled by 86-N force to the right. If the coefficient of friction is 0.19, find the acceleration of the system and the tension force between the masses.

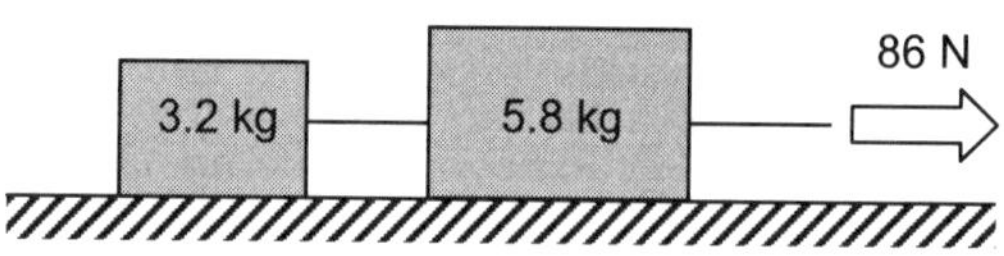

Copyright © by June Chung - All rights reserved. Reproduction of this publication without the written permission of the copyright owner is strictly prohibited.

ANSWER KEY TO EXAMPLES

Example 3-1

(a) The rock travels in mid-air after being dropped. Thus, it experiences only the force of gravity and accelerates at 9.8 m/s^2 towards the ground.

(b) The parked car experiences the force of gravity and the force by the ground. The forces are balanced, and the car does not accelerate.

(c) While being thrown, the baseball quickly accelerates in the direction of the applied force by the pitcher.

(d) The two forces are balanced. The object does not accelerate.

(e) The two forces cannot be balanced. The object does accelerate.

(f) The forces acting north and south cancel each other out, and there is no force to balance the force acting west. The object accelerates west.

Example 3-2

Since the eraser moves with constant velocity, the forces on it <u>must</u> be balanced. It implies that there must be a force acting upward to balanced the force of gravity ($\vec{F}_g$) and a force acting against the direction of motion to balance the applied force ($\vec{F}_A$). As you will learn later in this chapter, the push exerted by the surface ($\vec{F}_N$) balances the force of gravity, and the friction force ($\vec{F}_f$) balances the applied force. The diagram showing the forces acting on the eraser is drawn below:

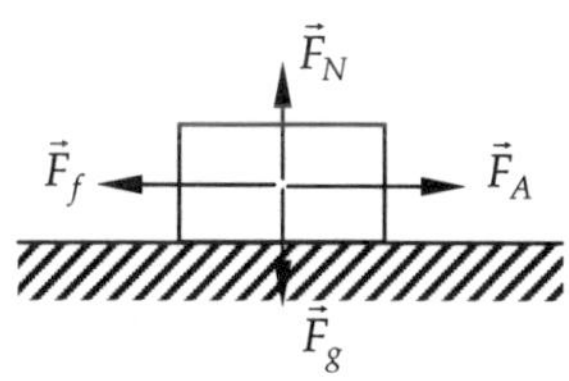

Example 3-3

The 25-N forces cancel each other. The sum of the horizontal forces is 28 N, left. Thus, the object's acceleration is

$$\vec{a} = \frac{\vec{F}_{net}}{m} = \frac{28}{5.2} = 5.4 \text{ m/s}^2\text{, left}$$

Example 3-4

Since Julie receives the 130-N force from behind, she accelerates according to Newton's second law:

$$\vec{a} = \frac{\vec{F}_{net}}{m} = \frac{130}{45} = 2.9 \text{ m/s}^2\text{, forward}$$

Copyright © by June Chung - All rights reserved. Reproduction of this publication without the written permission of the copyright owner is strictly prohibited.

According to Newton's third law, Jack receives the same amount of force in the opposite direction. Thus, his acceleration is obtained by

$$\vec{a} = \frac{\vec{F}_{net}}{m} = \frac{130}{60} = 2.2 \text{ m/s}^2, \text{ backward}$$

Example 3-5

(a) The weight of the dumbbell, $\vec{F}_g$, is obtained from

$$F_g = mg = 18(9.8) = 176.4 \text{ N, towards the ground}$$

The magnitude of the normal force is equal to that of the weight, which is 176.4 N. Thus, the normal force is 176 N, upward.

(b) Since the magnitude of the applied force is less than that of the weight, it is not sufficient to lift the dumbbell off the ground. Thus, the magnitude of the normal force decreases by 125 N. The normal force becomes 51 N, upward.

(c) This time the magnitude of the lifting force is large enough to lift the dumbbell off the ground. Since the dumbbell no longer makes contact with the ground, the normal force is nonexistent.

Example 3-6

The two masses are motionless, which implies that the forces on each mass are balanced. Mass m_2 is pulled upward by the lower string, and that tension force is balanced by the weight of the mass. As for m_1, it is acted upon by its own weight and two tension forces pulling it in opposite directions. The diagrams below show the forces on each mass:

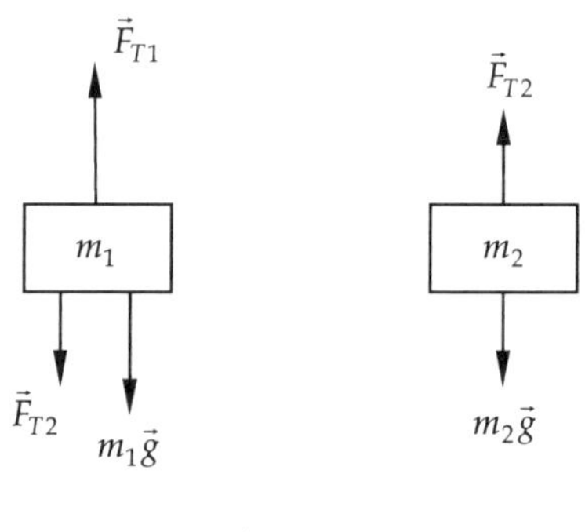

The tension force acting on m_2, $\vec{F}_{T2}$, has the same magnitude as the tension force pulling m_1 downward because the two masses are connected to the same string. From the diagrams, we find that

$$F_{T2} = m_2 g = (22)(9.8) = 220 \text{ N}$$

$$F_{T1} = F_{T2} + m_1 g = 220 + (15)(9.8) = 360 \text{ N}$$

Copyright © by June Chung – All rights reserved. Reproduction of this publication without the written permission of the copyright owner is strictly prohibited.

While the magnitude of the tension force in the lower string is equal to the weight of m_1, that of the tension force in the upper string is equal to the weight of the two masses combined.

Example 3-7

The three forces act on the object as shown below. When these force vectors are added using the tip-to-tail method, they form a right triangle where the angle between the gravitational force and the normal force is 30°. The magnitude of the gravitational force, which is

$$F_g = mg = (3.5)(9.8) = 34.3 \text{ N}$$

is the hypotenuse of the triangle. Thus, the magnitude of the friction is obtained by using trigonometry:

$$\sin 30° = \frac{\text{opposite}}{\text{hypotenuse}} = \frac{F_f}{F_g} = \frac{F_f}{34.3}$$

$$F_f = F_g \sin 30° = 34.3(0.50) = 17 \text{ N}$$

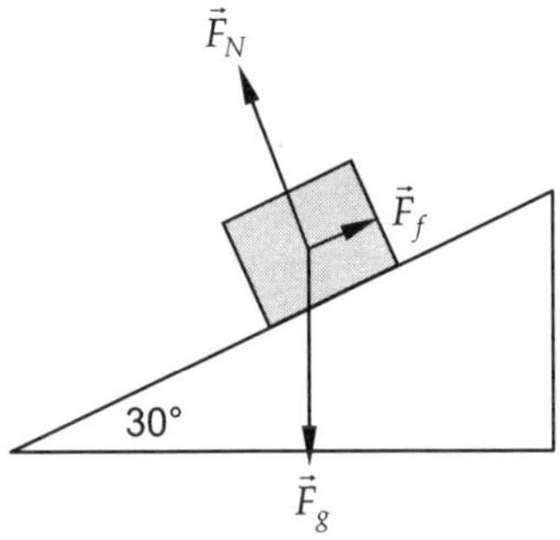

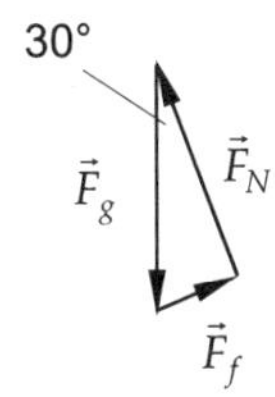

Example 3-8

Since the cart is moving at constant velocity, the forces on it must be balanced. The downward force of gravity, which is

$$F_g = mg = 32(9.8) = 314 \text{ N}$$

is balanced by the upward normal force, and the 56-N tension force is balanced by the friction force, which is

$$F_f = \mu F_N = \mu(314) = 56 \text{ N}$$

Therefore, the coefficient of friction is 0.18.

Example 3-9

(a) Draw the force of gravity vertically downward. Since the block is in contact with the surface, draw the normal force perpendicular to it and the friction force parallel to it. Since the block is moving up the incline, the friction force is directed down the incline. Finally, draw the tension force up the incline.

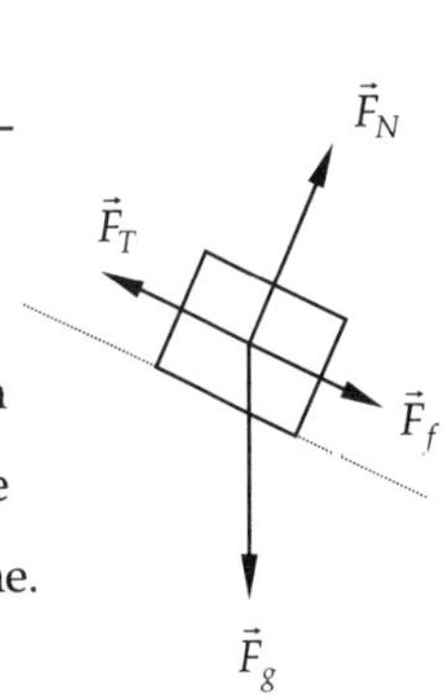

Copyright © by June Chung - All rights reserved. Reproduction of this publication without the written permission of the copyright owner is strictly prohibited.

(b) Isolate each mass and draw the force of gravity vertically downward in each diagram. Since both masses are in contact with the ground, draw the normal and friction forces for each. The mass on the left is also in contact with a string, so draw a tension force to the right. The mass on the right is in contact with two strings, one on each side, so draw tension forces accordingly.

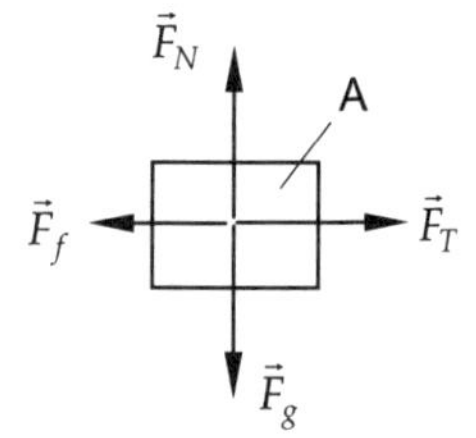

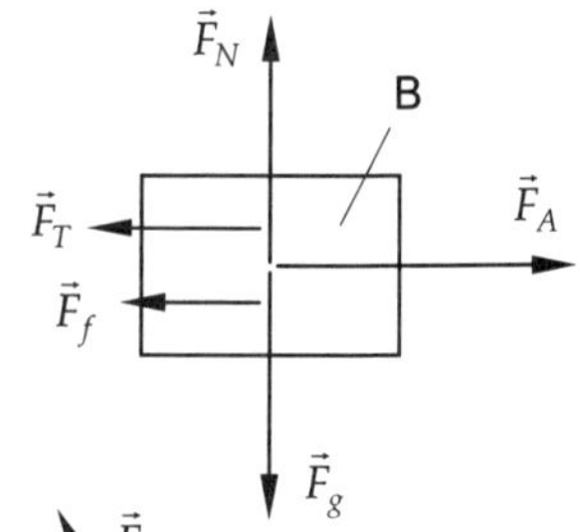

(c) Draw the force of gravity vertically downward. As the pendulum bob is in contact with the string, draw the tension force along the string, away from the bob. Finally, draw the applied force to the right. As the bob is motionless, the vector sum of these forces must be zero.

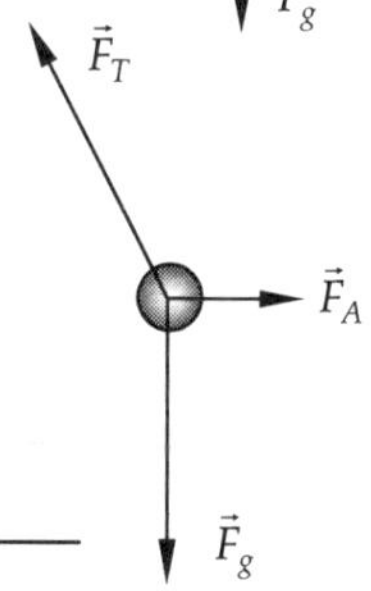

Example 3-10 Draw the free-body diagram of the cart as shown below:

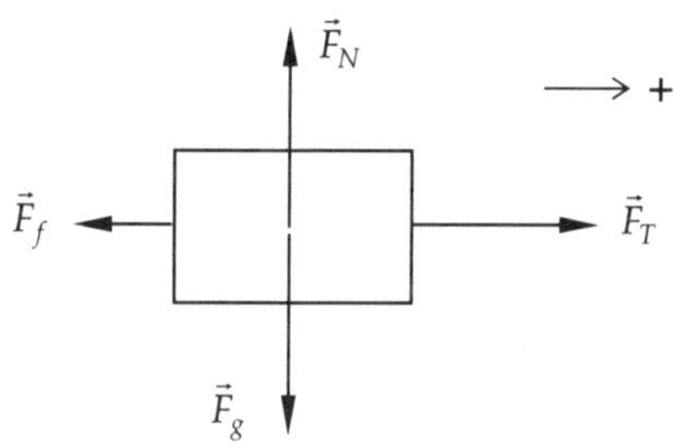

The positive direction is set to be to the right as the cart moves to right. Set the equation of motion as

$$F_T - F_f = m\,a$$

Substitute Eq. (3.3) for F_f:

$$F_T - \mu F_N = m\,a \qquad \dots (1)$$

The normal force can be found by equating it to the force of gravity:

$$F_N = F_g = m\,g = (32)(9.8) = 314\ \text{N}$$

Substitute the numbers into Eq. (1):

$$75 - 0.23\,(314) = 32\,a$$

Solve for a, and we find that the cart accelerates at 0.090 m/s^2 in the positive direction.

Copyright © by June Chung – All rights reserved. Reproduction of this publication without the written permission of the copyright owner is strictly prohibited.

Example 3-11

Draw the free-body diagram of each mass:

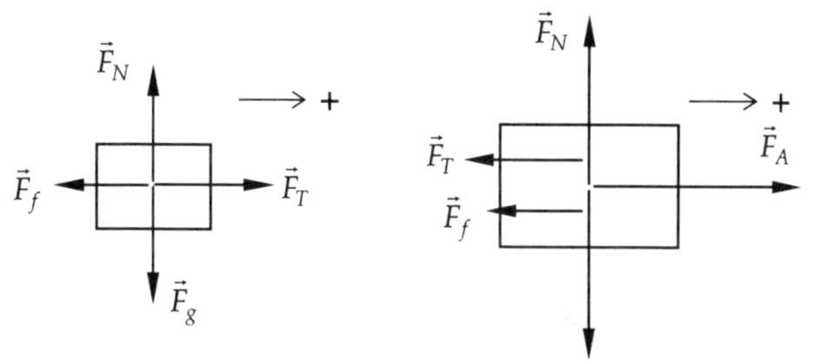

Set up the equation of motion for each mass:

3.2-kg mass: $F_T - \mu F_N = m a$

$$F_T - (0.19)(3.2)(9.8) = (3.2)a \qquad \dots (1)$$

5.8-kg mass: $F_A - F_T - \mu F_N = m a$

$$86 - F_T - (0.19)(5.8)(9.8) = (5.8)a \qquad \dots (2)$$

The above equations have two unknowns (the tension force and the acceleration). Solving them simultaneously, we obtain

$$a = 7.7 \text{ m/s}^2 \text{ and } F_T = 31 \text{ N}$$

Copyright © by June Chung – All rights reserved. Reproduction of this publication without the written permission of the copyright owner is strictly prohibited.

PROBLEMS ON CHAPTER 3: DYNAMICS

(Answer Key on p. 226)

1 - 10. Fill in the blanks

1. __________ is a push or a pull.
2. __________ is the tendency of an object to maintain its current state of motion.
3. The ________ ________ on an object is the vector sum of all the forces acting on it.
4. Newton's first law of motion is also called the law of __________.
5. The __________ of an object is the amount of material in it.
6. The __________ of an object is the force of gravity acting on it.
7. The __________ force is the force exerted by a surface perpendicular to the surface.
8. The __________ force is the force exerted by a surface parallel to the surface.
9. The __________ force is the force transmitted through a string.
10. A __________ __________ shows an object of interest and all the forces acting on it.

11 - 20. True/False questions

11. _____ If the forces on an object are balanced, it does not move.
12. _____ If an object is at rest, the forces on it are balanced.
13. _____ The acceleration of an object is inversely proportional to its mass.
14. _____ Newton's First Law of motion states that, "For every action, there is an equal and opposite reaction."
15. _____ The unit for weight is the kilogram [kg].
16. _____ For an object moving horizontally on level ground, the magnitude of the normal force on it is always equal to that of its weight.
17. _____ Normal forces always point away from the surface.
18. _____ Tension forces always point along the string, away from the object.
19. _____ An object at rest does not experience the friction force.
20. _____ The magnitude of the friction force depends on the speed of an object.

Copyright © by June Chung - All rights reserved. Reproduction of this publication without the written permission of the copyright owner is strictly prohibited.

21 - 28. Multiple choice questions

21. Which of the following statements is false for an object whose speed is changing?

    a) Its velocity is changing.
    b) Its acceleration is zero.
    c) Forces acting on the object are not balanced.
    d) At least one force is acting on the object.

22. When aboard a vehicle turning a sharp corner, you feel as if your body is being pushed out of the vehicle. Which Newton's law explains this phenomenon?

    a) Newton's First Law
    b) Newton's Second Law
    c) Newton's Third Law
    d) Newton's laws cannot explain it.

23. An object moving on level ground has a constant acceleration. Which of the following statements <u>must</u> be true?

    a) The object is experiencing friction force.
    b) An applied force is acting on the object.
    c) There are at least two forces acting on the object, and their sum is zero.
    d) There are at least three forces acting on the object, and their sum is not zero.

24. The free-body diagram of an object being pulled by a string on level ground is shown below. Which of the following forces is the reaction force to the force of gravity, $\vec{F}_g$?

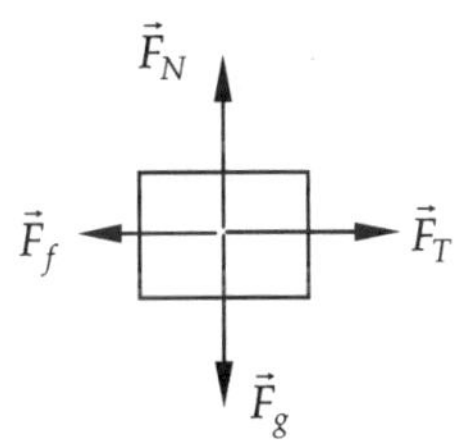

    a) the normal force, $\vec{F}_N$
    b) the tension force, $\vec{F}_T$
    c) the friction force, $\vec{F}_f$
    d) The reaction force is not shown in this diagram.

25. Object A weighs more than object B when they are weighed on an unknown planet. Which of the following is the correct statement?

    a) The mass of A is greater than the mass of B.
    b) The mass of B is greater than the mass of A.
    c) The mass of A is equal to the mass of B.
    d) The masses of the objects cannot be compared.

Copyright © by June Chung – All rights reserved. Reproduction of this publication without the written permission of the copyright owner is strictly prohibited.

26. Besides the coefficient of friction, the magnitude of the friction force on a moving object depends on

a) the object's weight
b) the normal force on the object
c) the object's speed
d) the net force on the object

27. An object is being pulled up a smooth inclined surface by a rope. Which of the following forces should not be included in the free-body diagram?

a) normal force
b) friction force
c) tension force
d) gravitational force

28. An object accelerates upward while being pulled by a rope. The tension in the rope is

a) greater than the object's weight
b) equal to the object's weight
c) less than the object's weight
d) zero

29 – 42. Written-response questions

29. A 150-kg object accelerates uniformly for 5.5 s from rest and travels 73 m. What is the magnitude of the net force on the object during its acceleration?

30. A 350-kg object collides head-on with a 66-kg object. During this collision, the 350-kg object accelerates backwards at 0.28 m/s^2. What is the acceleration of the 66-kg object?

31. Force *F*, when applied to object A, causes it to accelerate at 7.2 m/s^2. When the same force is applied to object B, it accelerates at 5.1 m/s^2. If the same force is applied to accelerate both objects together, what is their acceleration?

32. A string that can withstand the maximum tension force of 55 N is used to lift a 2.2-kg object. What is the maximum acceleration the object can attain without snapping the string?

33. A 29-N force is applied horizontally on a 7.9-kg object on level ground. If the object remains motionless despite the applied force, what is the magnitude of the friction force on the object? (The coefficient of friction between the ground and the object is 0.42.)

Copyright © by June Chung – All rights reserved. Reproduction of this publication without the written permission of the copyright owner is strictly prohibited.

34. Eric is riding an elevator to go from the ground floor to the 78th floor. Starting from rest, it accelerates uniformly and reaches the maximum speed of 7.2 m/s in 8.5 seconds. The elevator continues to move upward at the maximum speed, and when it is 25 m below the 78th floor, it begins to decelerate uniformly and comes to rest on the 78th floor. If Eric's weight is 580 N, find the magnitude of the normal force on him while (a) the elevator is accelerating and (b) the elevator is decelerating.

35. Two masses connected by a string are moving as shown. If there is no friction on the 8.3-kg object, find the magnitude of the tension force and the acceleration of the system.

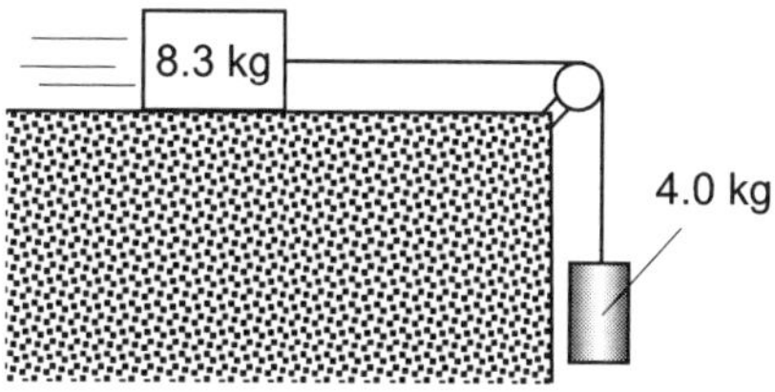

36. Jose throws a 1.50-kg object, initially at rest, vertically upward by applying a 22.0-N force through a displacement of 0.75 m.

a) Find the net force on the object while Jose applies the 22.0-N force.

b) Find how high the object rises after it leaves Jose's hand.

37. A 26-N force is applied horizontally to accelerate the two objects shown. If the coefficient of friction between the floor and the boxes is 0.47, what is the magnitude of the force exerted by the 2.5-kg object on the 1.5-kg object?

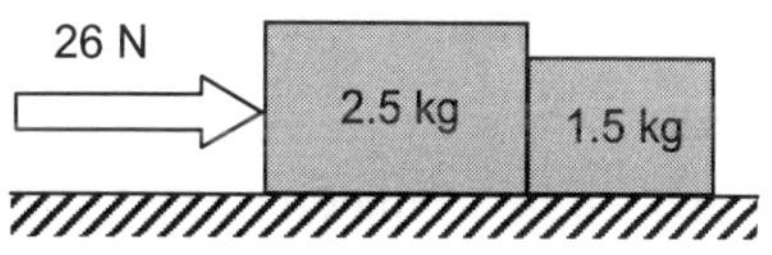

38. Two objects connected by a string are released from the position shown.

a) Find the acceleration of the objects and the tension in the string.

b) Determine time taken for the 19-kg object to reach the ground.

c) Find the velocity of the 11-kg object when the 19-kg object hits the ground.

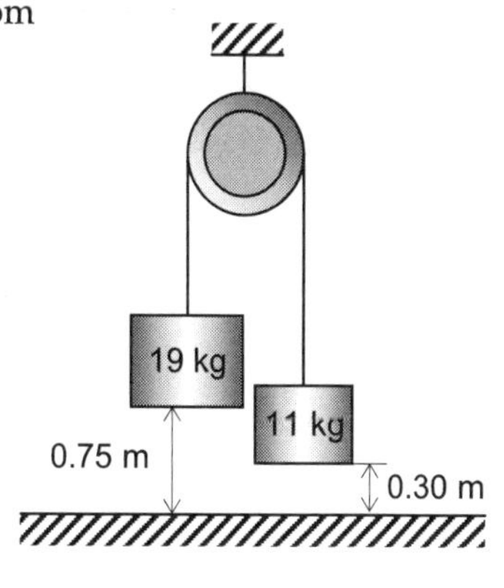

Copyright © by June Chung - All rights reserved. Reproduction of this publication without the written permission of the copyright owner is strictly prohibited.

39. A car of mass 1200 kg is traveling at a constant velocity of 90 km/h, north. The coefficient of friction between the tires and the road is 0.27.

    a) Draw a free-body diagram showing the magnitude and direction of all the forces acting on the car.

    b) Suddenly the car runs out of gas, but the driver sees a gas station 200 m along the highway. Will the car reach the station before coming to a halt? Show your calculations.

40. A 25-g bullet moving at high speed strikes a 1.5-kg block at rest. Upon impact, the bullet is embedded within the block, and they start moving at 13 m/s. If the coefficient of friction between the block and the ground is 0.45, how far does the block travel before coming to a stop? Show the free-body diagram of the block while it is moving. (Assume the block does not move until the bullet is fully embedded and comes to rest within it.)

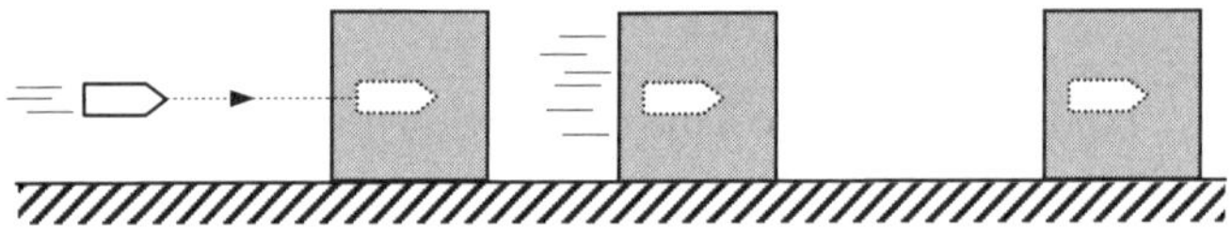

41. Reconsider Problem 40. If the same event occurred on the surface of the moon, where the gravitational field strength is 1.6 N/kg, would the block travel the same distance? If not, would it travel farther or shorter? Show your reasoning.

42. A 35-g bullet moving horizontally at 375 m/s strikes a wooden wall. Upon impact, the bullet slows down and stops 14 cm from the surface of the wall. Find the average force the wall exerts on the bullet. (The vertical motion of the bullet in this problem is negligible, so we can ignore the weight of the bullet and assume it travels horizontally until it comes to a full stop.)

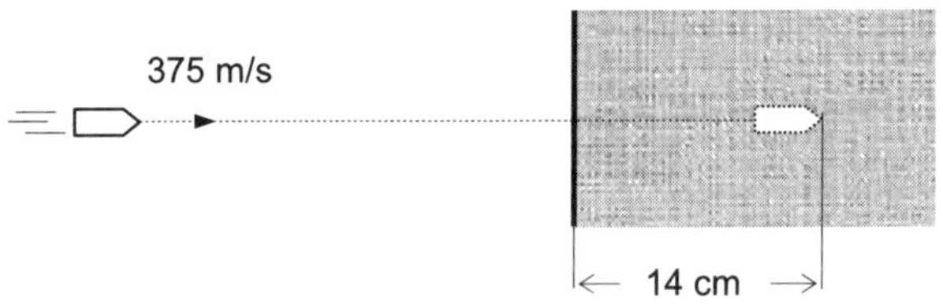

Copyright © by June Chung - All rights reserved. Reproduction of this publication without the written permission of the copyright owner is strictly prohibited.

CHAPTER 4

MOMENTUM

Momentum is what Newton called "the quantity" of motion. It describes the effect of a moving object. Indeed, he initially formulated his Laws of Motion using the concept of momentum and change in momentum. We shall meet the first conserved quantity, which is a new and important concept, and be introduced to the idea of an isolated system.

The importance of impulse in many sporting activities cannot be ignored. Why is Tiger Woods world-famous? Why are some tennis players better than others? Partly because they can hit the ball further and faster than their competitors, which means that they have learnt how to exert a force for a longer period of time. This imparts a greater impulse to the ball, which in turn gives it a greater velocity and range. Hence they become the top players.

In this chapter, you will study:

✓ momentum, change in momentum, and its relation to net force

✓ impulse and its relation to change in momentum

✓ the law of conservation of momentum

✓ how to solve collision and explosion problems

Copyright © by June Chung - All rights reserved. Reproduction of this publication without the written permission of the copyright owner is strictly prohibited.

Key Point 4.1: Momentum ($\vec{p}$)

momentum

Momentum is a vector quantity that refers to the quantity of motion an object possesses. It is defined as the product of mass and velocity:

$$\vec{p} = m\vec{v} \text{ [kg·m/s]} \qquad \dots (4.1)$$

The momentum vector is always in the direction of the motion. For example, if an object is moving east, its momentum is also to the east.

change in momentum

While an object's momentum can change when its mass or velocity changes, the change in momentum caused by a change in velocity is of particular interest and is written as

$$\Delta\vec{p} = m(\vec{v}_f - \vec{v}_i) = m\Delta\vec{v} \text{ [kg·m/s]} \qquad \dots (4.2)$$

Using the definition of acceleration in Eq. (2.4) (p. 27), we can express Eq. (4.2) as

$$\Delta\vec{p} = m\,\vec{a}\,\Delta t \qquad \dots (4.3)$$

relationship between change in momentum and net force

Since $\vec{F}_{net} = m\,\vec{a}$ according to Newton's second law of motion, we arrive at the following relationship between the change in momentum and the net force acting on the body:

$$\vec{F}_{net} = \frac{\Delta\vec{p}}{\Delta t} \text{ [kg·m/s}^2\text{]} \qquad \dots (4.4)$$

Therefore, the change in momentum is proportional to the net force, and the rate of change of momentum is the net force causing this change. If the net force on an object is zero, its momentum stays constant, as does its velocity. Note that the direction of the momentum is independent of that of the change in momentum, which has the same direction as the net force.

> - Momentum is the quantity of motion a body has.
> - The direction of the momentum is always the same as that of the velocity.
> - The greater the net force on a body, the greater the change in its momentum.
> - The direction of the change in momentum is independent of that of the momentum; the former has the same direction as the net force.

Copyright © by June Chung - All rights reserved. Reproduction of this publication without the written permission of the copyright owner is strictly prohibited.

Example 4-1

Liz, while throwing a 2.2-kg softball, accelerates the ball from rest to 13 m/s in 0.34 seconds. Calculate (a) the change in momentum of the ball and (b) the net force on the ball during the throw.

Example 4-2

Determine whether the momentum of each object remains constant.

(a) A ball thrown horizontally from the top of a tall building
(b) An airplane accelerating on a runway for take-off
(c) A racecar moving around a circular track at constant speed
(d) A crate sliding down a flat, 7° slope at constant speed

Copyright © by June Chung – All rights reserved. Reproduction of this publication without the written permission of the copyright owner is strictly prohibited.

Key Point 4.2: Impulse ($\vec{J}$)

impulse

Impulse is a vector quantity and is defined as the product of the net force applied to an object and the time period over which the force was applied:

$$\vec{J} = \vec{F}_{net}\,\Delta t \quad [\text{N·s} = \text{kg·m/s}] \qquad \ldots (4.5)$$

Impulse always points in the same direction as the net force. Thus, if an object receives an upward net force, the impulse on it is also directed upward. For example, when a free-falling object hits the ground, the object receives an upward force during impact while the ground receives a downward force. (These forces are an action-reaction pair.) Therefore, the impulse on the object is upward, and that on the ground is downward.

relationship between impulse and change in momentum

Using Eq. (4.4), we can also define impulse in terms of the change in momentum as

$$\vec{J} = \vec{F}_{net}\,\Delta t = \Delta\vec{p} = m\left(\vec{v}_f - \vec{v}_i\right) \qquad \ldots (4.6)$$

This is a very useful equation because you can obtain impulse from net force and time or from mass and velocities. How you obtain impulse depends on which set of quantities are known.

> - Impulse is the product of the net force on a body and the time the force acts.
> - Direction of impulse is always the same as that of the net force on a body.
> - Impulse causes the change in momentum and is equal to it.

Example 4-3

A 2.6-kg ball is thrown horizontally towards a solid wall at 11 m/s. After the impact, the ball bounces back at the same speed. Has the momentum of the ball changed? If so, what is the impulse on the ball?

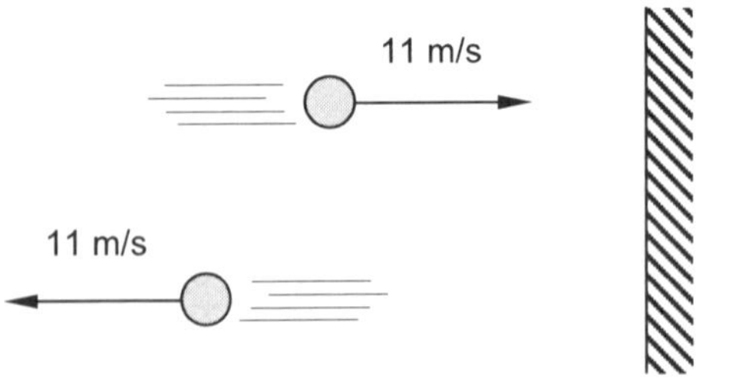

Copyright © by June Chung - All rights reserved. Reproduction of this publication without the written permission of the copyright owner is strictly prohibited.

Key Point 4.3: **The Law of Conservation of Momentum**

In a collision between two masses or in an explosion of an object into two smaller pieces, both masses experience forces which are equal in magnitude but opposite in direction, according to Newton's third law of motion. For instance, suppose two objects, *A* and *B*, collide with each other, as shown in Fig. 4-1. Then we have

derivation of the principle of conservation of momentum

$$\vec{F}_{on\,A\,by\,B} = -\vec{F}_{on\,B\,by\,A} \qquad \ldots (4.7)$$

As the time of contact, Δt, is the same for *A* and *B*,

$$\vec{F}_{on\,A\,by\,B} \cdot \Delta t = -\vec{F}_{on\,B\,by\,A} \cdot \Delta t \qquad \ldots (4.8)$$

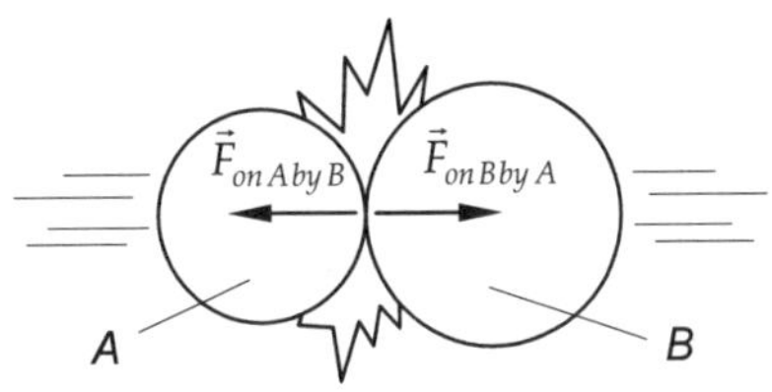

Figure 4-1: Interaction between colliding masses

$\vec{F} \cdot \Delta t$ is defined as impulse, which is also equal to the change in momentum, $\Delta \vec{p}$. Thus, we can re-write Eq. (4.8) as

$$\Delta \vec{p}_{of\,A} = -\Delta \vec{p}_{of\,B} \qquad \ldots (4.9)$$

or

$$\left(\vec{p}_{of\,A}\right)_{after} - \left(\vec{p}_{of\,A}\right)_{before} = -\left[\left(\vec{p}_{of\,B}\right)_{after} - \left(\vec{p}_{of\,B}\right)_{before}\right] \qquad \ldots (4.10)$$

The subscripts 'before' and 'after' indicate the motions before and after the collision, respectively. By re-arranging the terms in Eq. (4.10), we finally obtain the equation for the **conservation of momentum** for a system of two masses:

$$\left(\vec{p}_{of\,A} + \vec{p}_{of\,B}\right)_{before} = \left(\vec{p}_{of\,A} + \vec{p}_{of\,B}\right)_{after} \qquad \ldots (4.11)$$

or

$$\left(m_A \vec{v}_A + m_B \vec{v}_B\right)_{before} = \left(m_A \vec{v}_A + m_B \vec{v}_B\right)_{after} \qquad \ldots (4.12)$$

Hence, the sum of the momenta of colliding objects does not change as a result of a collision or explosion. In fact, in the absence of any external force, the sum of the momenta in any system of masses is always conserved. In other words, when no external force acts on a system, the

the law of conservation of momentum

Copyright © by June Chung - All rights reserved. Reproduction of this publication without the written permission of the copyright owner is strictly prohibited.

total momentum of the system before the interaction is equal to the total momentum afterwards. Consequently, when you observe objects colliding or exploding, you should immediately think of conservation of momentum and apply the equation for it. (In this course, you may always assume that no external force is involved.)

> - Whenever you see a collision or an explosion...
>
>   ⟹ Think conservation of momentum!
>
> For <u>all</u> types of collisions and explosions,
>
> - Use $(m_A\vec{v}_A + m_B\vec{v}_B)_{before} = (m_A\vec{v}_A + m_B\vec{v}_B)_{after}$

Example 4-4

Determine whether the sum of their momenta is conserved during the following interaction between objects.

(a) A moving billiard ball hits a stationary billiard ball.

(b) An astronaut throws a tool away while floating in space.

(c) Two cars collide head-on in a crash test.

Copyright © by June Chung - All rights reserved. Reproduction of this publication without the written permission of the copyright owner is strictly prohibited.

Key Point 4.4: Collisions and Explosions

perfectly inelastic collisions

While most colliding objects are in contact with each other only during the collision and separate afterwards, some objects become locked during the collision and move away together afterwards. This is called a **perfectly inelastic collision**. An example is when a bullet is fired into a block of wood, gets embedded in it, and moves with the block. For such collisions, we can reduce Eq. (4.12) to

$$(m_A\vec{v}_A + m_B\vec{v}_B)_{before} = (m_A + m_B)\vec{v}_{after} \qquad \ldots (4.13)$$

because both objects move with the same velocity after the collision.

explosions

You can think of an **explosion** as the reverse of a perfectly inelastic collision. Exploding objects are initially locked together before the explosion and separate afterwards. The equation for an explosion can be written as

$$(m_A + m_B)\vec{v}_{before} = (m_A\vec{v}_A + m_B\vec{v}_B)_{after} \qquad \ldots (4.14)$$

Keep in mind that Eqs. (4.12), (4.13), and (4.14) are all vector equations. Therefore, you need to be careful about the direction of the velocities.

- In a perfectly inelastic collision, colliding objects stick and move together after the collision.
- For such collisions, the law of conservation of momentum can be written as
$$(m_A\vec{v}_A + m_B\vec{v}_B)_{before} = (m_A + m_B)\vec{v}_{after}$$
- Explosions are the reverse of perfectly inelastic collisions. The equation of conservation of momentum for explosions is
$$(m_A + m_B)\vec{v}_{before} = (m_A\vec{v}_A + m_B\vec{v}_B)_{after}$$

Copyright © by June Chung – All rights reserved. Reproduction of this publication without the written permission of the copyright owner is strictly prohibited.

Key Point 4.5: **How to Solve Collision and Explosion Problems**

Problems involving collisions and explosions require you to use the principle of conservation of momentum. As long as you take care with the direction of the velocities, they are generally easy to solve. The solution procedure is outlined in four steps:

1. Read the problem statement carefully. Make sure two objects are colliding or two objects initially joined together are separating.
2. Draw a diagram showing both objects before and after the collision (or before and after the separation). An example is shown below.

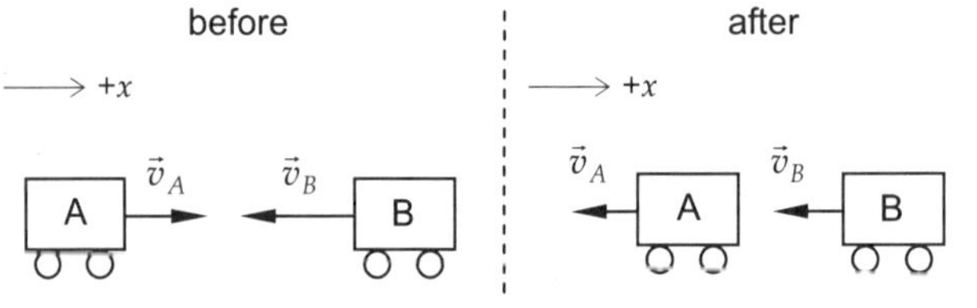

3. Choose the positive direction, and calculate the momentum of each object before and after the collision (or before and after the separation) based on that direction.
4. Set up Eq. (4.12). Check that there is only one unknown in the equation. Solve for the unknown, and think about whether the answer makes sense.

Example 4-5

A 5.3-kg object, initially traveling east at 3.5 m/s, collides with a 8.1-kg object at rest. If they stick and travel together after the collision, what is their velocity? What impulse is given to the 8.1-kg object?

Example 4-6

Mary and her little brother, Nathan, were skating together at 4.8 m/s when Nathan suddenly pushes Mary so that her speed increases to 6.1 m/s. If Mary and Nathan respectively weigh 430N and 350 N, find Nathan's speed after the push. (Assume Mary and Nathan travel in a straight line.)

Copyright © by June Chung - All rights reserved. Reproduction of this publication without the written permission of the copyright owner is strictly prohibited.

ANSWER KEY TO EXAMPLES

Example 4-1

(a) Set the direction of the throw as positive. Then the ball's initial momentum is zero and its final momentum is $(2.2)(13) = +29$ kg·m/s. The change in momentum is then $+29 - 0 = +29$ kg·m/s, or 29 kg·m/s in the direction of the throw.

(b) According to Eq. (4.2), the net force can be found by dividing the change in momentum by the time of the throw:

$$\vec{F}_{net} = \frac{\Delta \vec{p}}{\Delta t} = \frac{29}{0.34} = 84 \text{ N in the direction of the throw}$$

(Alternatively, you may calculate the ball's acceleration first and multiply it with its mass to obtain the net force.)

Example 4-2

Only the crate in (d) is moving with constant velocity. The ball in (a) accelerates towards the ground, and the velocity of the racecar in (c) is not constant because it continually changes the direction of motion. Since the velocity changes when an object accelerates, only those moving at constant velocity maintain constant momentum. Hence, the momentum is constant only for the crate in (d).

Example 4-3

Since the direction of the ball changes after the impact, there is a change in its velocity, which means its momentum changes as well. We can obtain the impulse by finding the change in momentum. First, let's assume the direction of the ball's initial velocity as positive. Then the initial velocity becomes +11 m/s and the final velocity -11 m/s. Thus,

$$\Delta \vec{p} = m\Delta \vec{v} = (2.6)(-11 - 11) = -57 \text{ kg·m/s}$$

The negative sign indicates the direction. So the impulse on the ball is 57 kg·m/s, away from the wall.

Example 4-4

The sum of the interacting objects' momenta is conserved in all cases.

Example 4-5

This is a perfectly inelastic collision. The diagram showing both objects before and after the collision is shown below:

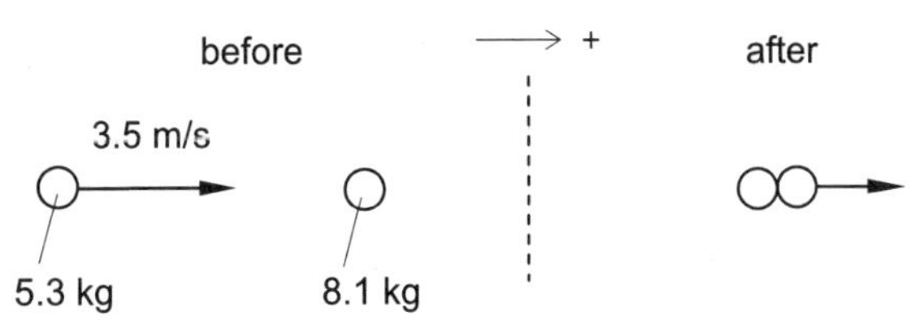

Copyright © by June Chung – All rights reserved. Reproduction of this publication without the written permission of the copyright owner is strictly prohibited.

Set east as positive, and calculate the momentum of each object before and after the collision:

5.3-kg object: $\vec{p}_{before} = (5.3)(3.5) = 18.6\ \text{kg·m/s}$

$\vec{p}_{after} = (5.3)\vec{v}_{after}$

8.1-kg object: $\vec{p}_{before} = (8.1)(0) = 0\ \text{kg·m/s}$

$\vec{p}_{after} = (8.1)\vec{v}_{after}$

Set up the equation for conservation of momentum:

$$18.6 + 0 = 5.3\,\vec{v}_{after} + 8.1\,\vec{v}_{after} = 13.4\,\vec{v}_{after}$$

Solve for the unknown, and we obtain

$$\vec{v}_{after} = 1.4\ \text{m/s, east}$$

The impulse given to the 8.1-kg object is equal to the change in its momentum:

$$\vec{J} = \vec{p}_{after} - \vec{p}_{before} = (8.1)(1.4) - 0 = 11\ \text{N·s, east}$$

Example 4-6

This is an explosion problem. The diagram showing Mary and Nathan before and after the separation is shown below:

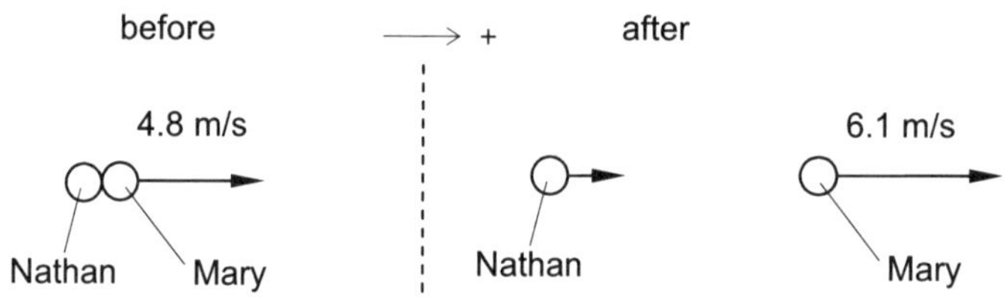

Set the initial direction of motion as positive, and calculate the momentum of Mary and Nathan before and after the separation:

Mary: $\vec{p}_{before} = \left(\frac{430}{9.8}\right)(4.8) = 211\ \text{kg·m/s}$

$\vec{p}_{after} = \left(\frac{430}{9.8}\right)(6.1) = 268\ \text{kg·m/s}$

Nathan: $\vec{p}_{before} = \left(\frac{350}{9.8}\right)(4.8) = 171\ \text{kg·m/s}$

$\vec{p}_{after} = \left(\frac{350}{9.8}\right)\vec{v}_{after}$

Set up the equation for conservation of momentum:

$$211 + 171 = 268 + \left(\frac{350}{9.8}\right)\vec{v}_{after}$$

We find that $\vec{v}_{after} = 3.2\ \text{m/s}$, in the initial direction of motion.

Copyright © by June Chung – All rights reserved. Reproduction of this publication without the written permission of the copyright owner is strictly prohibited.

PROBLEMS ON CHAPTER 4: MOMENTUM

(Answer Key on p. 229)

1 - 4. Fill in the blanks

1. __________ is the amount of motion an object possesses and is defined as the product of the object's mass and __________.

2. The ________ ________ on an object indicates how fast its momentum changes with respect to time.

3. __________ is defined as the product of net force applied to an object and time period over which the force was applied. It is equal to the change in __________.

4. In _______ ________ collisions, colliding masses stick together during the collision.

5 - 12. True/False questions

5. _____ The momentum of an object is always in the direction of its motion.

6. _____ The greater the momentum of a body, the greater the net force on it.

7. _____ When an object's speed changes, its momentum changes as well.

8. _____ When an object receives a non-zero impulse, its momentum changes.

9. _____ If there is no force acting on an object, its momentum remains constant.

10. _____ When an object's momentum is constant, it travels in a straight line.

11. _____ When two objects collide, the sum of their momenta is conserved only if the they stick together during the collision.

12. _____ When a stationary object explodes into two pieces, the pieces must travel in opposite directions.

13 - 18. Multiple choice questions

13. The momentum of an object is constant if __________________. Which of the following makes the above statement false?

    a) the net force on the object is zero.
    b) the acceleration of the object is zero.
    c) the impulse on the object is zero.
    d) the change in the object's speed is zero.

Copyright © by June Chung – All rights reserved. Reproduction of this publication without the written permission of the copyright owner is strictly prohibited.

14. An egg does not break when dropped onto a mattress while it does when dropped from the same height onto a tiled floor. This occurs because

   a) the egg experiences the greater change in momentum when dropped onto the tiled floor.

   b) the egg receives the greater average net force when dropped onto the tiled floor.

   c) the egg has the greater impact speed when dropped onto the tiled floor.

   d) All of the above

15. A 3.0-kg object initially heading south collides with another object and turns east. Which of the following could be the direction of the impulse received by the 3.0-kg object?

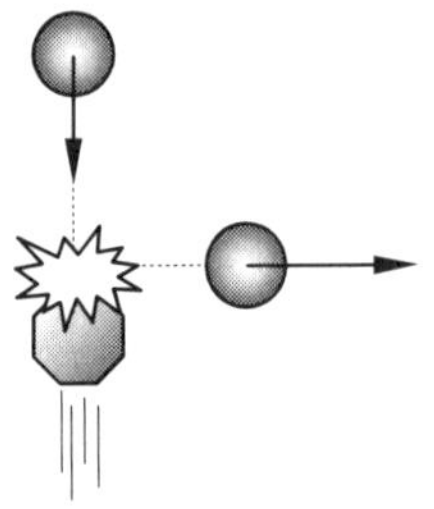

   a) 

   b) 

   c) 

   d) 

16. A train and a motorcycle collide head-on. Which of the following is true?

   a) The motorcycle receives the greater magnitude of impulse.

   b) They receive the same magnitude of impulse.

   c) The momentum of the train is conserved.

   d) The motorcycle sticks to the train.

17. When you fire a gun at a target in front of you, the gun recoils back toward you. Why is the gun's speed less than that of the bullet?

   a) Because the gun receives less magnitude of impulse.

   b) Because the gun has less magnitude of momentum.

   c) Because the gun has the greater mass.

   d) Because the gun receives less magnitude of force.

Copyright © by June Chung - All rights reserved. Reproduction of this publication without the written permission of the copyright owner is strictly prohibited.

18. An object initially flying east is split into two pieces, as shown. If the first piece head directly upward after the separation, which of the following could be the direction of the other piece's momentum?

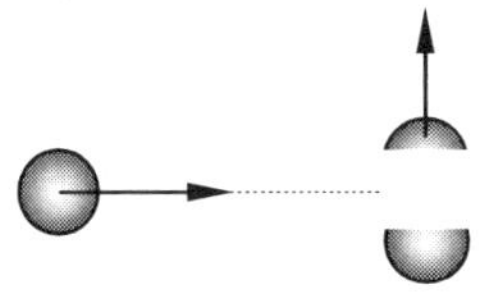

a) 

b) 

c) 

d) 

19 - 31. Written-response questions

19. A 28-N object is thrown vertically upward at 14 m/s. What is the object's momentum after 1.8 seconds?

20. A 15-kg object initially at rest receives a 30-N force for 3.0 seconds. What is the final speed of the object?

21. A 7.3-kg object is traveling on level ground at 16 m/s. What is the minimum average net force needed to stop this object within 0.75 seconds?

22. Object A has twice the mass of object B. When two objects are at rest, we apply a 25-N force to object A for 5.0 seconds. If we want object B to attain the same speed as object A, how long should we apply a 10-N force to object B?

23. An applied force causes a 9.5-kg object to accelerate uniformly from rest. If this object moves 5.5 meters in 2.0 seconds, how much impulse has the applied force exerted on the object?

24. An object initially moving at 2.1 m/s, north, collides head-on with another object heading south and receives an average force of 54 N for 0.33 seconds. If the velocity of the first object changes to 1.8 m/s, north, after the collision, what is its mass?

Copyright © by June Chung - All rights reserved. Reproduction of this publication without the written permission of the copyright owner is strictly prohibited.

25. A tennis ball of mass 110 g hits a wall traveling at 14.2 m/s. If it rebounds directly backwards at 13.8 m/s and the impact lasts for 0.15 s, what is the average net force that the wall exerts on the ball?

26. Water is ejected from a hose at a rate of 3.9 kg/s. It hits a wall at 13 m/s and is splashed back at the same speed. What is the magnitude of the force exerted by the water on the wall?

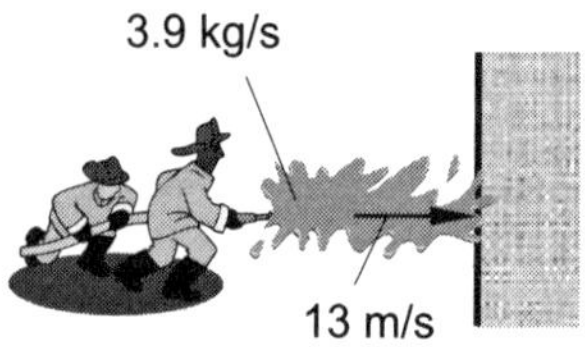

27. A 16-kg cart was initially moving at 4.8 m/s on a horizontal surface when a 4.5-kg bag of cement is suddenly dumped into the cart. What is the speed of the cart loaded with the bag?

28. A missile with a mass of 1500 kg is moving north at 420 m/s when it suddenly explodes into eleven pieces. What is the sum of the pieces' momenta?

29. A 0.12-kg ball moving north at 30 cm/s collides with a 0.22-kg ball moving south. After the collision, the speed of each ball is reduced by half, and the 0.12-kg ball changes the direction of its motion. (The 0.22-kg ball continues to move south after the collision.) What was the initial speed of the 0.22-kg ball?

30. A 6.0-kg object initially at rest explodes into two pieces. The smaller piece, whose mass is 2.0-kg, has the velocity of 7.5 m/s, west. The bigger piece soon explodes again and divides into a 1.0-kg piece and a 3.0-kg piece. If the 1.0-kg piece comes to a complete stop after the explosion, what is the velocity of the 3.0-kg piece?

31. A 25-g bullet traveling horizontally at 350 m/s strikes a 1.5-kg block at rest. Upon impact, the bullet is embedded within the block. If the coefficient of friction between the block and the ground is 0.45, how far does the block travel before coming to a stop? (Assume the block does not move until the bullet is fully embedded and comes to rest within it.)

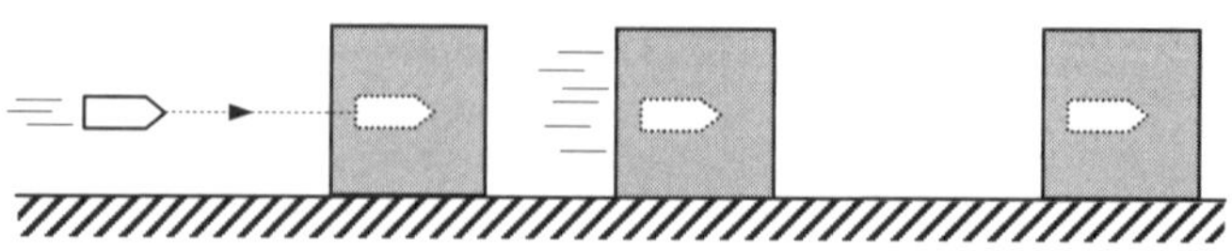

Copyright © by June Chung – All rights reserved. Reproduction of this publication without the written permission of the copyright owner is strictly prohibited.

CHAPTER 5

WORK, ENERGY, AND POWER

At last we have reached our objective. We can now quantify energy in terms of force and displacement. In physics, doing work has a very precise meaning. Studying these notes and reading your textbook while preparing for a test does not constitute "work" in the physical sense, although moving your pen over the paper to write notes and do examples does qualify as work. When you do physical work, you use some of your energy. Indeed, it is impossible to do work without using energy, and both work and energy are measured in Joules. Work, not love, is what makes the world go around because work is the way that energy is transferred from one object to another, and conservation of energy underpins all change and is the basis of Physics.

In this chapter, you will study:

- ✓ energy and its conservation
- ✓ work and its relation to energy
- ✓ kinetic energy and its relation to work
- ✓ potential energy
- ✓ mechanical energy and its conservation
- ✓ how to solve work-energy problems
- ✓ power and efficiency

Copyright © by June Chung - All rights reserved. Reproduction of this publication without the written permission of the copyright owner is strictly prohibited.

Key Point 5.1: Energy and Its Conservation

conservation of energy

Energy is a very useful quantity because it is conserved during an interaction. Indeed, the **law of conservation of energy** is the corner-stone of physics. This law states that energy can be neither created nor destroyed; it can merely change from one form to another. So, if you know the total amount of energy inside any isolated system† before an interaction, it remains the same afterwards.

example

For example, suppose a ball is dropped inside an isolated chamber. As the ball bounces on the floor, it keeps losing energy, eventually coming to rest. In the meantime, the molecules within the chamber have acquired the ball's initial energy, converted into heat, sound energy, and so forth. Hence, if we define our system to be the chamber and the space within it, the total amount of energy within the system before the ball is dropped is equal to that at any instant afterwards.

conservation of energy within the universe

In the most extreme case, we can even apply this concept of energy conservation to our universe. Even though all kinds of energy exchanges occur among the constituent particles within the universe, they are of no interest to us when we define our universe to be the system. Instead, we should focus on work or heat crossing the boundary of our system. Assuming that no such energy exchanges occur, we can then state that the total amount of energy within the universe remains unchanged.

> ➢ The law of conservation of energy:
> "The total amount of energy within an object or a system of objects is conserved if there is no energy exchange between it and its surroundings."

† An isolated system is one in which no energy is transferred to or from its surroundings.

Copyright © by June Chung - All rights reserved. Reproduction of this publication without the written permission of the copyright owner is strictly prohibited.

Key Point 5.2: **Work (W)**

finding work done by a force

When a force causes an object to move, we say that the force is doing **work** on the object. If the object moves in the same direction as the force, the work done on the object by the force is given by

$$W = F\,d \quad [\text{N·m} = \text{kg·m}^2/\text{s}^2 = \text{J}] \qquad \dots (5.1)$$

where F is the magnitude of the individual force (not the net force) and d is the magnitude of the object's displacement. (Note that work is a scalar quantity, and it is measured in Joules [J].) On the other hand, if the object moves in the opposite direction to the force, the work done on the object is

$$W = -F\,d \qquad \dots (5.2)$$

Finally, if the object moves perpendicular to the direction of the force, the work done on the object is zero. In any case, if the object does not move, the work done on it is always zero.

example

Now, suppose a 30-N box is being pulled parallel to the ground by a 20-N force at constant velocity, as shown in Fig. 5-1. The term "constant velocity" tells you that the forces on the box are balanced. In this case, the normal force balances the weight, and the friction force balances the 20-N applied force. If the box has moved 5.0 m in the direction of the pull, you can calculate the work done on the box by each force as:

$$W_{by\ applied\ force} = F_A d = 20(5.0) = 100\ \text{J}$$

$$W_{by\ friction\ force} = -F_f d = -20(5.0) = -100\ \text{J}$$

$$W_{by\ weight} = W_{by\ normal\ force} = 0\ \text{J}$$

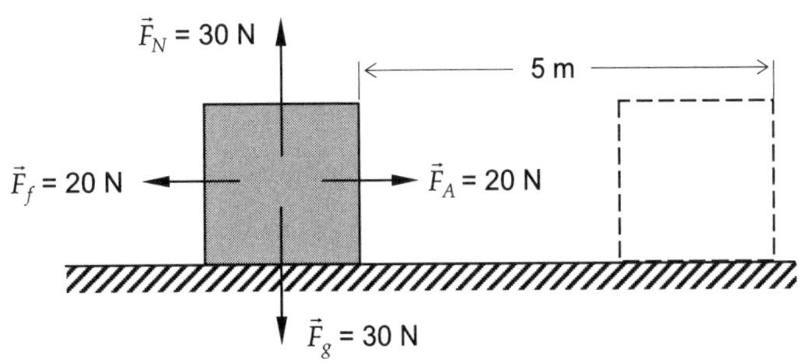

Figure 5-1: Work done on a mass by various forces

Copyright © by June Chung – All rights reserved. Reproduction of this publication without the written permission of the copyright owner is strictly prohibited.

net work

To find the net work done on the box, simply add up all the work done by the individual forces. In the above example, the net work is zero. In fact, whenever the net force on an object is zero, the net work on it is also zero.

relationship between work and energy

The significance of work done on an object is that it indicates a transfer of energy into or out of the object. If a force does positive work on an object, the force transfers its energy into the object, and the level of energy within the object increases. If it does negative work on an object, it draws energy out of the object, and the level of energy within the object decreases. If a force does not do any work on an object, there is no transfer of energy into or out of the object.

Let's reconsider the box in the above example. Observing the amount of work done by each force, we can figure out that 100 J of energy has been delivered by the applied force into the object, but the same amount of energy has been taken out of the object in the form of heat energy resulting from friction. As a result, the net gain in energy by the object during its motion is zero.

> - Work indicates a transfer of energy into or out of a body.
>
> If the angle between the force on a body and its displacement is...
>
> - 0°, the force does positive work of $W = Fd$
>   ⟹ Energy into the object!
> - 90°, the force does no work
> - 180°, the force does negative work of $W = -Fd$
>   ⟹ Energy out of the object!

Example 5-1

A 3.0-kg object is being pulled to the right by a 36-N force and moves 2.1 m along the surface. If the coefficient of friction is 0.44, find the work done by each force acting on the object, and the net work done on it.

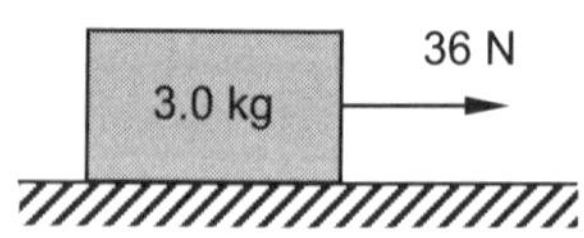

Copyright © by June Chung – All rights reserved. Reproduction of this publication without the written permission of the copyright owner is strictly prohibited.

Key Point 5.3: **Kinetic Energy (E_K)**

derivation of formula for kinetic energy

The net work done on an object is expressed as

$$W_{net} = F_{net} d = m\,a\,d \quad [\text{kg·m}^2/\text{s}^2 = \text{J}] \qquad \dots (5.3)$$

Substituting Eq. (2.8) (p. 33) for *d* into Eq. (5.3), we obtain

$$W_{net} = m\,a\left(v_i t + \frac{1}{2}at^2\right) = m\,a\,t\left(v_i + \frac{1}{2}at\right) \qquad \dots (5.4)$$

According to Eq. (2.4) (p. 27),

$$a\,t = v_f - v_i \qquad \dots (5.5)$$

Substitute Eq. (5.5) into Eq. (5.4). After simplifying the resulting equation, we find that the net work done on an object is given by

$$W_{net} = \frac{1}{2}m\left(v_f^2 - v_i^2\right) \quad [\text{kg·m}^2/\text{s}^2 = \text{J}] \qquad \dots (5.6)$$

Thus, the net work needed to be done on an object of mass *m* kg to accelerate it from rest to *v* m/s is

kinetic energy

$$\frac{1}{2}m\,v^2 = E_K \qquad \dots (5.7)$$

which we define to be the **kinetic energy** of the object. Being proportional to the square of the speed, it can never become negative.

Using Eq. (5.7), we can re-write Eq. (5.6) to show the relationship between work and kinetic energy:

work-energy principle

$$W_{net} = \Delta E_K \qquad \dots (5.8)$$

This equation is known as the **work-energy principle**. It is one of the most important concepts in this chapter, and it states that the net work done on an object causes a change in its kinetic energy. If positive net work is done on an object, it gains energy, which is used to speed up the object. If negative work is done on an object, the object releases part of its energy and loses speed, resulting in a decrease in its kinetic energy.

> - Kinetic energy is the energy due to motion.
> - A change in kinetic energy of a body is caused by the net work done on it.
>
>   Positive net work ⟹ increase in kinetic energy
>
>   Negative net work ⟹ decrease in kinetic energy

Copyright © by June Chung - All rights reserved. Reproduction of this publication without the written permission of the copyright owner is strictly prohibited.

Example 5-2

Tammy wants to accelerate a 15-kg object from rest to 4.2 m/s while pushing it 7.5 m on a frictionless, horizontal surface. How much force does she need to exert on the object?

Example 5-3

A 3.7-kg object was moving at 8.9 m/s when a net work of -120 J is done on it. What is the new speed of the object, v_f? How much net work must now be applied to double v_f?

Copyright © by June Chung – All rights reserved. Reproduction of this publication without the written permission of the copyright owner is strictly prohibited.

Key Point 5.4: Potential Energy (E_P)

potential energy

Potential energy is stored energy associated with the position of a body within the field of another body. There are many kinds of potential energy such as gravitational, elastic, and electric potential energy. In Physics 11, we learn about gravitational potential energy.

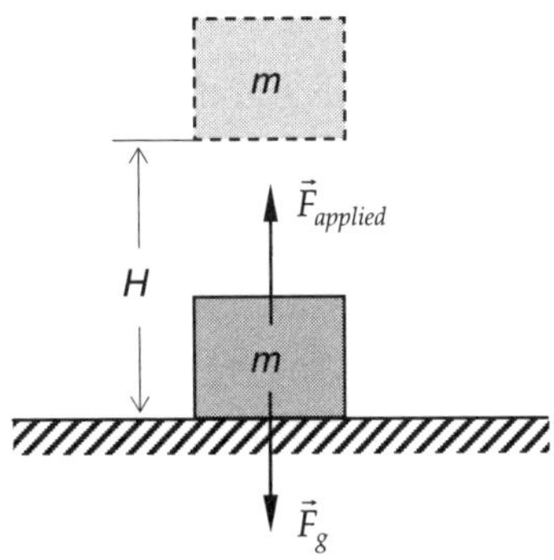

Figure 5-2: Work done by an applied force to raise an object

derivation of formula for gravitational potential energy

Suppose an object initially on the ground is raised vertically upward at constant velocity by an applied force, as shown in Fig. 5-2. According to Newton's first law, the magnitude of the applied force must equal that of the object's weight, mg. Then the amount of work done on the object by the applied force as the object moves up a vertical distance H is

$$W = F_{Applied}\,H = m\,g\,H \quad [\text{kg}\cdot(\text{m/s}^2)\cdot\text{m} = \text{J}] \quad \ldots (5.9)$$

This work brings about a change in the object's potential energy, and the gravitational potential energy is defined as

$$E_P = m\,g\,h \quad \ldots (5.10)$$

where h is the vertical height of the object above some reference level, at which $h = 0$. Then we can express Eq. (5.9) as

$$W = m\,g\,H = \Delta E_P = m\,g\,(h_f - h_i) \quad \ldots (5.11)$$

choosing a reference level

Note that it does not matter where you set the reference level. Just choose a convenient level, and make all your calculations from that level. In the example above, if we choose the ground to be the reference level, the object's gravitational potential energy becomes zero at the initial position and mgH at the final position. However, if we choose the final altitude as the reference level, the gravitational potential energy

Copyright © by June Chung - All rights reserved. Reproduction of this publication without the written permission of the copyright owner is strictly prohibited.

becomes $-mgH$ at the initial position and zero at the final position. Notice that we have a negative gravitational potential energy at the initial position! But that is fine. The actual value of the energy, whether it be kinetic or potential, is not what is really important. It is the <u>change</u> in the energy that counts. In fact, the change in gravitational potential energy is the same regardless of your choice of reference level, as shown below:

Reference at the initial position: $\Delta E_P = mgH - 0 = mgH$

Reference at the final position: $\Delta E_P = 0 - (-mgH) = mgH$

> - Potential energy is the stored energy associated with the position of a body within the field of another body.
> - Gravitational potential energy is defined as *mgh*, where *h* is the vertical height above the reference level; the reference level may be set arbitrarily.
> - The actual value of energy is not what counts; it is the <u>change in energy</u> that counts.

Example 5-4

A 2.5-kg ball is dropped from the top of a 75-m building and hits the ground below. Find the change in the ball's gravitational potential energy, setting the top of the building to be the reference level.

Example 5-5

An object falls to the ground along the following paths. Compare the change in the object's gravitational potential energy along different paths.

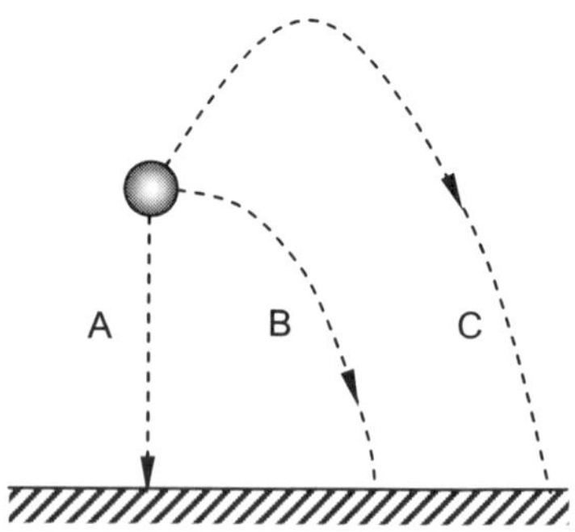

Copyright © by June Chung – All rights reserved. Reproduction of this publication without the written permission of the copyright owner is strictly prohibited.

Key Point 5.5: Conservation of Mechanical Energy

mechanical energy

The **mechanical energy** of an object is defined as the <u>sum</u> of its potential and kinetic energies. In the absence of any friction or applied force, the mechanical energy of an object is conserved during its motion:

$$(E_P + E_K)_{initial} = (E_P + E_K)_{final} \qquad \dots (5.12)$$

law of conservation of mechanical energy

Equation (5.12) is known as the **law of conservation of mechanical energy**. Using Eqs. (5.7) and (5.10), we can express Eq. (5.12) as

$$mgh_i + \frac{1}{2}mv_i^2 = mgh_f + \frac{1}{2}mv_f^2 \qquad \dots (5.13)$$

Divide both sides of the equation by $m/2$:

$$2gh_i + v_i^2 = 2gh_f + v_f^2 \qquad \dots (5.14)$$

Finally, after re-arranging the terms, we obtain the following equation:

$$v_f^2 = v_i^2 + 2g(h_i - h_f) \qquad \dots (5.15)$$

comparison with one kinematic equation

Equation (5.15) looks familiar, doesn't it? Compare it to Eq. (2.9) in Key Point 2.9 (p. 33), and you realize that they are actually the same equation. The only difference is that Eq. (5.15) assumes no friction or applied force while there was no such restriction when we derived Eq. (2.9). If a friction force or an applied force exists, the mechanical energy of an object may not be conserved, and you must use the work-energy principle in Eq. (5.8) to analyze the motion of the object.

> - Mechanical energy of a body is the sum of its potential and kinetic energies.
> - Mechanical energy is conserved in the absence of friction and applied forces.
> - The law of conservation of mechanical energy:
>
> $$(E_P + E_K)_{initial} = (E_P + E_K)_{final}$$
>
> or $$v_f^2 = v_i^2 + 2g(h_i - h_f)$$

Example 5-6

A 250-g object is thrown horizontally from the top of a 65-m building with the initial speed of 28 m/s. (a) What is the speed of the object when it is halfway to the ground? (b) How much has the object fallen when its speed reaches 42 m/s?

Copyright © by June Chung - All rights reserved. Reproduction of this publication without the written permission of the copyright owner is strictly prohibited.

Key Point 5.6: Perfectly Elastic Collisions

perfectly elastic collision

In Chapter 4 we learned that, when two objects interact in a collision, the sum of their momenta is conserved. However, the sum of their kinetic energies may or may not be conserved in a collision. **A perfectly elastic collision** is one in which kinetic energy is conserved. This kind of collision is very rare and only occurs between smooth, hard spherical objects and between molecules of a substance at a given temperature. Usually, part of the masses' kinetic energy is dissipated during a collision and goes into the surroundings in the form of heat, sound, and sometimes light energy. Thus you should never assume that a collision is perfectly elastic, but if you are told that it is, then set up the following equation

$$(E_{KA} + E_{KB})_{before} = (E_{KA} + E_{KB})_{after} \quad \text{... (5.16)}$$

and use it with the equation for the conservation of momentum in Eq. (4.12) to solve problems.

Collisions are classified into...

1. perfectly elastic collisions: The sum of the colliding masses' kinetic energies is conserved as well as the sum of their momenta.
2. inelastic collisions: Only the sum of the colliding masses' momenta is conserved. The masses separate after collision.
3. perfectly inelastic collisions: Only the sum of the colliding masses' momenta is conserved. The masses lock and move together after colliding.

Example 5-7

A 35-kg object traveling at 5.0 m/s, west, collides with a 15-kg object moving at 12.0 m/s, east, and heads back east at 4.3 m/s after the collision. Determine whether the collision is perfectly elastic. If not, calculate how much of the objects' initial kinetic energies was lost during the collision.

Copyright © by June Chung – All rights reserved. Reproduction of this publication without the written permission of the copyright owner is strictly prohibited.

Key Point 5.7: How to Solve Work-Energy Problems

Most problems involving work and energy can be solved easily if you correctly identify the initial and final states of motion. If a friction or applied force is involved, apply the work-energy principle. Otherwise, apply conservation of mechanical energy. The solution procedure is summarized in four steps listed below:

1. Read the problem statement carefully and identify the initial and final states of motion. Draw a diagram to show the starting point, the journey, and the final point, and fill in the data on the diagram. Do not jump to the next step until you are sure what the initial and final states are.
2. Choose an appropriate point to be your reference height. Write down everything you know about the initial and final states such as height and speed. If there is no mention of height in the problem, assume the object is moving horizontally. Also, write down the mass and the displacement of the object if they are known.
3. Observe whether a friction force or applied force is involved. If any one of them is, apply the work-energy principle and set up Eq. (5.8). If not, use conservation of mechanical energy and set up Eq. (5.13) or Eq. (5.15).
4. Substitute the numbers into the equation. Check that there is only one unknown in the equation. Solve for the unknown and compare your answer with your own experience. For instance, you expect an object sliding on level ground to slow down because of friction and a falling object to gain speed in the absence of an applied force.

Copyright © by June Chung – All rights reserved. Reproduction of this publication without the written permission of the copyright owner is strictly prohibited.

Example 5-8

A block is initially moving up an inclined surface with the speed of 4.7 m/s. If the surface of the slope is smooth, how far does the block travel before coming back down the slope?

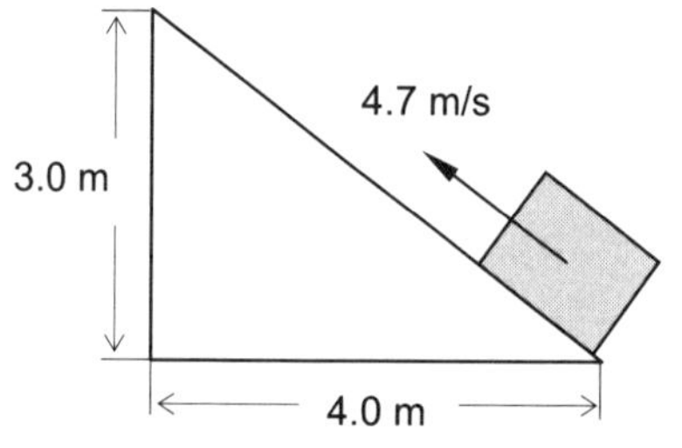

Example 5-9

An 8.3-kg object slides across a rough surface with an initial speed of 19 m/s. If the coefficient of friction is 0.35, how far does the object travel before its speed is cut in half?

Copyright © by June Chung – All rights reserved. Reproduction of this publication without the written permission of the copyright owner is strictly prohibited.

Key Point 5.8: Power (P) and Efficiency (η)

power

Power is a scalar quantity and is defined as the rate of doing work, or as the rate at which energy is transferred. It is expressed mathematically as

$$P = \frac{W}{t} = \frac{F\,d}{t} = F\,v \ \ [\mathrm{W} = \mathrm{J/s}] \qquad \ldots (5.17)$$

Here, you need to use the force exerted by an object, not the force acting on it, if you want to calculate the power delivered by the object. Hence, if you do one Joule of work per second, you are generating a power of one Watt [W].

kW·h

Note that the kW·h is not a unit of power although power companies sell electricity by the kilowatt-hour. (After all, it is not power that 'power' companies are selling to their customers!) In fact, the kW·h is a unit of energy and is equivalent to 3.6×10^6 Joules. Power companies use the kW·h because it is a larger unit and easier to deal with than the Joule when calculating energy usage in homes and businesses. For example, if we use a 100-W light bulb for 10 hours, how much energy did we consume? Answering in Joules requires some calculations; it is easier to answer it in kW·h because

$$100\,W\times(10\times3600)s = 3.6\times10^6\ \mathrm{J}$$

$$100\,W\times10\,h = 1000\,W\cdot h = 1\,kW\cdot h$$

So if you are told that you used 1000 kW·h last month, or about 33 kW·h per day, you would have an idea about the magnitude of your energy consumption.

efficiency

One of the ways to reduce your electricity usage at home is to use more efficient appliances. **Efficiency** of a machine is defined as

$$\eta = \frac{P_{out}}{P_{in}} = \frac{W_{out}}{W_{in}} \qquad \ldots (5.18)$$

where P_{in} is the power input to the machine and P_{out} is the power output of the machine. If you are given two numbers for P_{in} and P_{out} and are not sure whether the efficiency is $P_{in}\,/\,P_{out}$ or $P_{out}\,/\,P_{in}$, remember that efficiency is always less than or equal to 1. Hence, it is always the smaller number divided by the larger one. For example, if you supply 100 W of power to a machine, it can only generate 100 W or

Copyright © by June Chung – All rights reserved. Reproduction of this publication without the written permission of the copyright owner is strictly prohibited.

less. If it could generate more than 100 W, it would mean that you could successively amplify the amount of power to near infinity by connecting a number of such machines in series. That certainly sounds wonderful, but it is a physically impossible scenario as this violates the law of conservation of energy.

> - Power is the rate at which work is done or energy is transferred:
>
> $$P = \frac{W}{t} = F\,v$$
>
> - The kW·h is a unit of energy and is equivalent to 3.6×10^6 Joules.
> - Efficiency = power out / power in
> - Efficiency is always less than or equal to 1.

Example 5-10

A machine can lift a 90-kg object from the ground to the top of a 23-m building in 30 seconds. If its efficiency is 45%, determine the power output of the machine and the power input to the machine.

Example 5-11

Pamela has a 400-W machine that is 47% efficient, and she uses it for 8.0 hours everyday. If she replaces the machine with a newer model that requires the same power input but is 63% efficient, determine (a) the number of hours per day Pamela uses the newer model to perform the same task and (b) how much money she will save in four weeks (assuming it costs six cents to use 1 kW·h of energy).

Copyright © by June Chung - All rights reserved. Reproduction of this publication without the written permission of the copyright owner is strictly prohibited.

ANSWER KEY TO EXAMPLES

Example 5-1

First, draw the free-body diagram of the object as shown. The normal force balances the force of gravity, which is

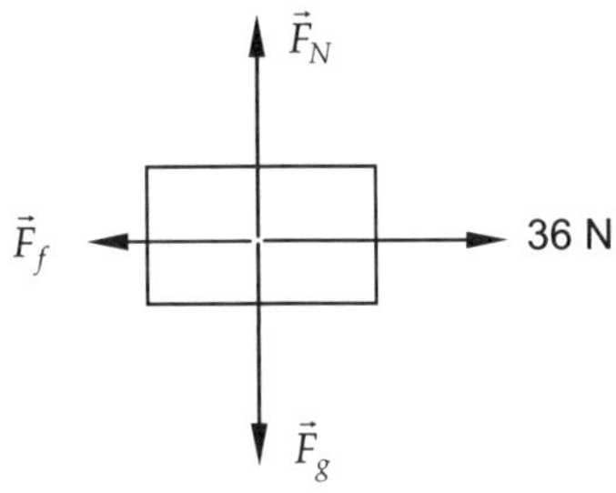

$$m\,g = (3.0)(9.8) = 29.4 \text{ N}$$

The magnitude of the friction force is therefore

$$F_f = \mu F_N = (0.44)(29.4) = 12.9 \text{ N}$$

Calculate the work done by each force:

Work done by $\vec{F}_g$ and $\vec{F}_N$ = 0 J because $\vec{F}_g$ and $\vec{F}_N \perp \vec{d}$.

Work done by the 36 N force = $36(2.1) = 76$ J

Work done by $\vec{F}_f$ = $-(12.9)(2.1) = -27$ J (It is negative because $\vec{F}_f$ is in the opposite direction of the displacement.)

Finally, the net work done on the object is 76 + (-27) = 49 J.

Example 5-2

Use Eq. (5.6) to calculate the amount of net work required to accelerate the object from rest to 4.2 m/s:

$$W_{net} = \frac{1}{2}(15)\left(4.2^2 - 0^2\right) = 132 \text{ J}$$

The net work on the object is the product of the force Tammy exerts on the object and the object's displacement:

$$W_{net} = F_{applied}(7.5) = 132 \text{ J}; \quad F_{applied} = 18 \text{ N}$$

Thus, Tammy needs to push the object with an 18-N force.

Example 5-3

Use Eq. (5.6) to find v_f:

$$-120 = \frac{1}{2}(3.7)\left(v_f^2 - 8.9^2\right)$$

The new speed of the object is therefore 3.8 m/s.

When the speed is doubled to 7.6 m/s, the amount of work needed is

$$W = \frac{1}{2}(3.7)\left(7.6^2 - 3.8^2\right) = 80 \text{ J}$$

Example 5-4

Since h = 0 at the top, the initial E_P is zero. The final height is 75 m below the top of the building, so h = -75 m. The final E_P is then $(2.5)(9.8)(-75) = -1800$ J. Thus, ΔE_P = -1800 J. (Now, set the ground to be the reference level and confirm that we still obtain ΔE_P = -1800 J.)

Copyright © by June Chung – All rights reserved. Reproduction of this publication without the written permission of the copyright owner is strictly prohibited.

Example 5-5

Near the surface of the earth, the change in gravitational potential energy is expressed as

$$\Delta E_P = m g \left(h_f - h_i\right)$$

Hence, when an object moves from point P to Q, the change in its gravitational potential energy depends only on the change in height. The route by which the object goes from P to Q does not affect ΔE_P. Since the object in this example goes through the same changes in height, the change in its gravitational potential energy is identical no matter which path it follows.

Example 5-6

(a) While the object is moving in mid-air, it experiences no friction or applied force. Hence, its mechanical energy is conserved until it hits the ground. Setting the ground to be the reference level, we can find the total mechanical energy of the object at the top of the building:

$$E_K + E_P = \frac{1}{2}(0.250)(28)^2 + (0.250)(9.8)(65) = 257 \text{ J}$$

When the object is halfway to the ground, its mechanical energy is still 257 J:

$$\frac{1}{2}(0.250)v_f^2 + (0.250)(9.8)(32.5) = 257 \text{ J}$$

$$v_f = 38 \text{ m/s}$$

(b) The object's total mechanical energy is still 257 J when its speed reaches 42 m/s:

$$\frac{1}{2}(0.250)(42)^2 + (0.250)(9.8)h_f = 257 \text{ J}$$

$$h_f = 15 \text{ m}$$

Thus, the object has fallen 50 m when its speed reaches 42 m/s. (Note that we can also use Eq. (5.15) to find solutions to (a) and (b).)

Example 5-7

The velocity of the 15-kg object after the collision can be obtained by using the principle of conservation of momentum.

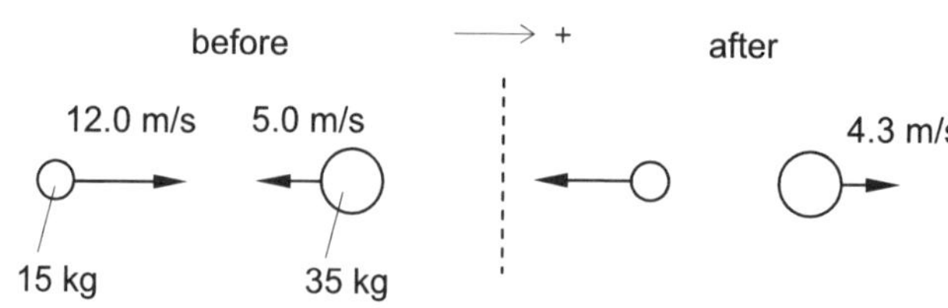

Copyright © by June Chung – All rights reserved. Reproduction of this publication without the written permission of the copyright owner is strictly prohibited.

First, the diagram above shows the objects before and after the collision. Apply Eq. (4.12):

$$(15)(12.0)+(35)(-5.0)=(15)\vec{v}_f+(35)(4.3)$$

Solving for $\vec{v}_f$, we find that the speed of the 15-kg object after the collision is 9.7 m/s. Thus, the sum of the kinetic energies before the collision is

$$\frac{1}{2}(35)(5.0)^2+\frac{1}{2}(15)(12.0)^2=1520 \text{ J}$$

while the sum of the kinetic energies after the collision is

$$\frac{1}{2}(35)(4.3)^2+\frac{1}{2}(15)(9.7)^2=1030 \text{ J}$$

Because the sum of the kinetic energies is not conserved, the collision is not perfectly elastic, and 490 J of energy is lost during the collision.

Example 5-8

The initial state of motion is when the block is moving at the speed of 4.7 m/s, and the final state is when the block is momentarily stopped at the maximum height.

Let h = 0 at the initial state, and write down the values of all known quantities:

$$v_i=4.7 \text{ m/s},\ v_f=0 \text{ m/s},\ h_i=0 \text{ m}$$

Since the surface is smooth, there is no friction force. Nor is there an applied force on the block. Hence, the mechanical energy of the block is conserved. At the initial state, the block's mechanical energy is

$$E_K+E_P=\frac{1}{2}m(4.7)^2+m(9.8)(0)=11m \text{ J}$$

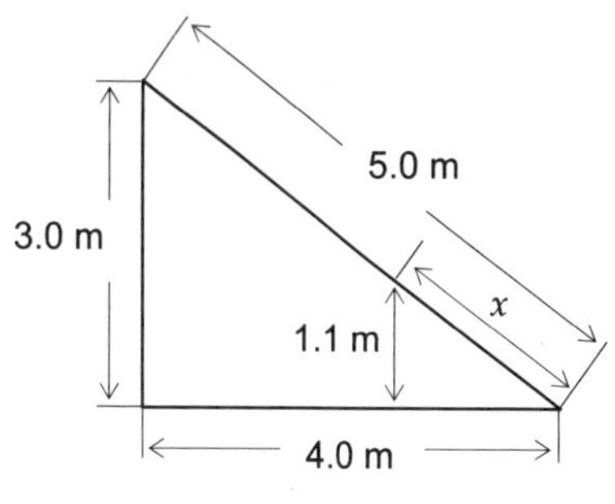

At the final state, it is

$$E_K+E_P=\frac{1}{2}m(0)^2+m(9.8)h_f=9.8mh_f \text{ J}$$

Set them equal to each other and solve for h_f:

$$h_f=1.13 \text{ m}$$

Find the distance of travel, x, by using the fact that x : 1.1 = 5.0 : 3.0. Solving for x, we find that the block travels 1.9 m up the slope before coming back down.

Copyright © by June Chung - All rights reserved. Reproduction of this publication without the written permission of the copyright owner is strictly prohibited.

Example 5-9

The initial state of motion is when the object is moving at the speed of 19 m/s, and the final state is when its speed is reduced to 9.5 m/s.

Since there is no mention of height, we assume the ground is level. Write down the values of all known quantities:

$$v_i = 19 \text{ m/s}, \; v_f = 9.5 \text{ m/s}, \; h_i = h_f = 0 \text{ m}$$

Because there is friction force on the object, apply the work-energy principle, and set up Eq. (5.6). To find the net work on the object, first draw the free-body diagram as shown right (assuming the object is moving to the right). Then,

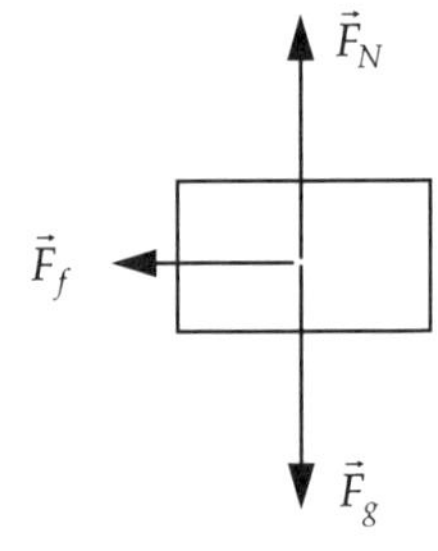

$$W_{by\ weight} = W_{by\ normal\ force} = 0 \text{ J}$$

$$W_{by\ friction\ force} = -F_f d = -\mu F_N d = -0.35(8.3)(9.8)d = -28.5\,d \text{ J}$$

Substitute the numbers into Eq. (5.6):

$$-28.5\,d = \frac{1}{2}(8.3)(9.5^2 - 19^2)$$

Solving for *d*, we find that the object travels 39 m before coming to a stop.

Example 5-10

The work needed to lift a 90-kg object 23 m above the ground is

$$W_{by\ machine} = F_A d = m g d = (90)(9.8)(23) = 20300 \text{ J}$$

Thus, the power output of the machine is

$$P_{out} = \frac{20300}{30} = 676 \text{ W}$$

Using Eq. (5.18), we can find the power input to the machine:

$$\eta = 0.45 = \frac{676}{P_{in}} \implies P_{in} = 1500 \text{ W}$$

Example 5-11

(a) The old machine consumed $(400)(8.0) = 3.2$ kW·h of energy while delivering 47% of 3.2 kW·h, which is 1.5 kW·h. Suppose the new machine is operated for *x* hours to do the same work of 1.5 kW·h. Then we can set up the following equation and solve for *x*:

$$(400)x(0.63) = 1500 \text{ W·h}; \quad x = 6.0 \text{ hours}$$

(b) In one day, the new machine consumes $(400)(6.0) = 2.4$ kW·h and saves 0.8 kW·h of energy. In four weeks, the total energy saved becomes 22.4 kW·h, which, when converted into dollars, is

$$22.4\,kW\cdot h \times \frac{6\,cents}{1\,kW\cdot h} = \$1.34$$

Copyright © by June Chung – All rights reserved. Reproduction of this publication without the written permission of the copyright owner is strictly prohibited.

PROBLEMS ON CHAPTER 5: WORK, ENERGY, & POWER (Answer Key on p. 230)

1 - 7. Fill in the blanks

1. _________ done on an object indicates the transfer of energy into or out of it.
2. The _______ _______ states that the net work done on an object is equal to the change in its kinetic energy.
3. _________ _________ energy is stored energy associated with the position of a body within the gravitational field of another body.
4. The mechanical energy of an object is the sum of its _____ _____ and _____ _____.
5. A _________ _________ collision is one in which the sum of the colliding masses' kinetic energies is conserved.
6. _______ is defined as the rate of doing work, or as the rate at which energy is transferred.
7. The _________ of a machine is defined as the power output over the power input.

8 - 16. True/False questions

8. _____ If a force is being applied to an object moving perpendicular to the force, the work done by the force on the object is zero.
9. _____ The work done by the friction force on a moving object is always positive.
10. _____ The net work done on an object is zero if the forces on it are balanced.
11. _____ An object's speed remains constant when the net work done on it is zero.
12. _____ An object above the ground has a positive gravitational potential energy relative to the ground.
13. _____ An object has a constant gravitational potential energy when the net work on it is zero.
14. _____ Mechanical energy of a free-falling object is conserved.
15. _____ If one machine can generate more power than another, its efficiency is higher than the efficiency of the other.
16. _____ The kilowatt-hour is a unit of energy and is equivalent to 3600 kJ.

Copyright © by June Chung – All rights reserved. Reproduction of this publication without the written permission of the copyright owner is strictly prohibited.

17 – 24. Multiple choice questions

17. An object initially at rest is accelerated uniformly. Which of the following statements could be false?

    a) The object received a constant force during the acceleration.
    b) Positive work was done on the object during the acceleration.
    c) The object's kinetic energy increased during the acceleration.
    d) The object's mechanical energy increased during the acceleration.

18. An object is thrown vertically upward from the ground. Which of the following graphs best represents the object's potential and kinetic energy during the flight?

a) 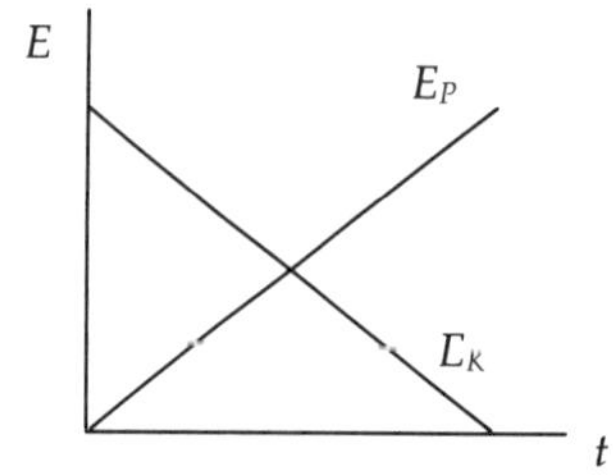

b) 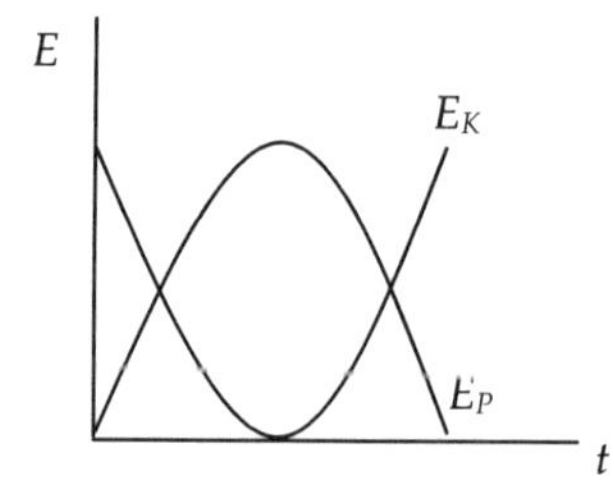

c) 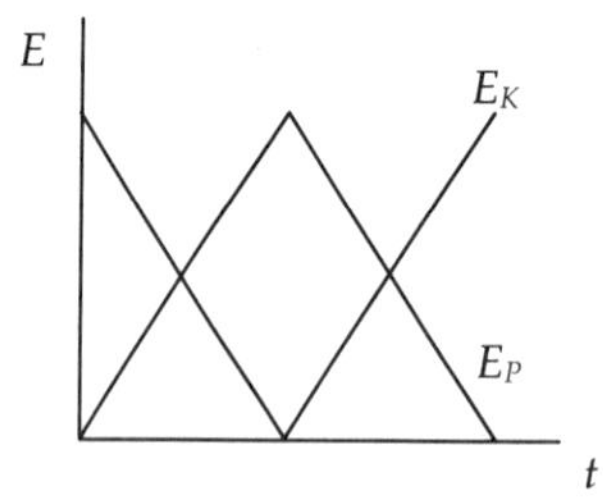

d) 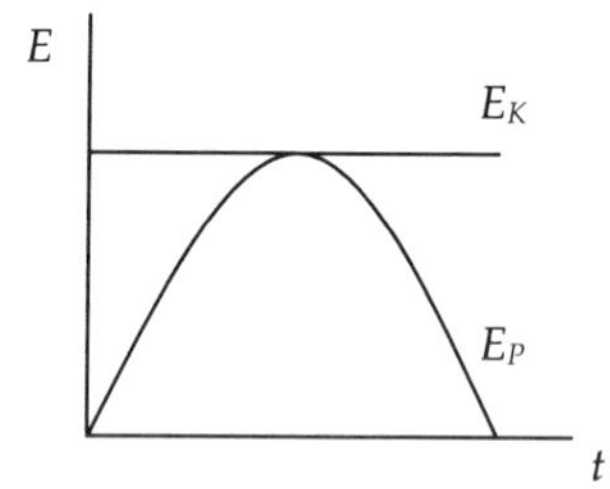

19. Tom and Jerry, two very good skiers, starts from the top of two 25-m high hills, as shown. For Tom, the surface of the downhill course is straight. For Jerry, it is steep at the beginning and gradually flattens near the bottom. With negligible friction, who will have the greater speed at the bottom of the hill?

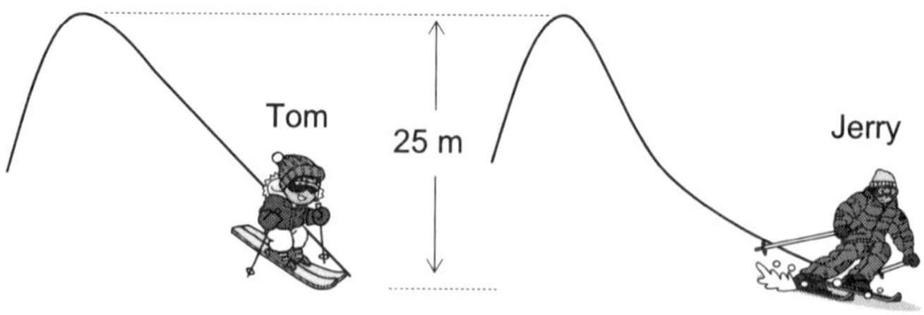

a) Tom
b) They both have the same speed.
c) Jerry
d) It is impossible to compare their speeds.

Copyright © by June Chung - All rights reserved. Reproduction of this publication without the written permission of the copyright owner is strictly prohibited.

20. A box is sliding down an inclined surface at constant speed as shown. Which of the following statements is not true?

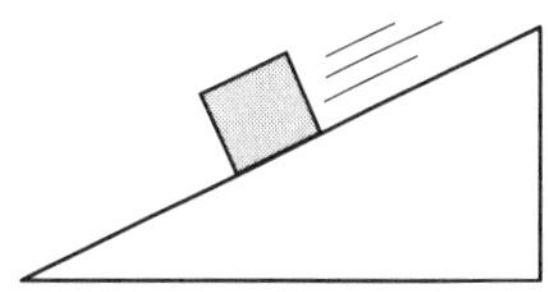

a) The work done by the normal force is zero.
b) The work done by the friction force is negative.
c) The net work done on the box is positive.
d) The gravitational potential energy of the box is decreasing.

21. Which of the following statements is true for a perfectly elastic collision between two objects?

a) The sum of their momenta is conserved.
b) One of the objects is at rest before the collision.
c) The kinetic energy of each object remains constant.
d) The colliding objects move with the same speed after the collision.

22. Three balls are thrown with the same initial speed but at different angles from the top of a tall building as shown. Which ball has the greatest speed at impact with the ground?

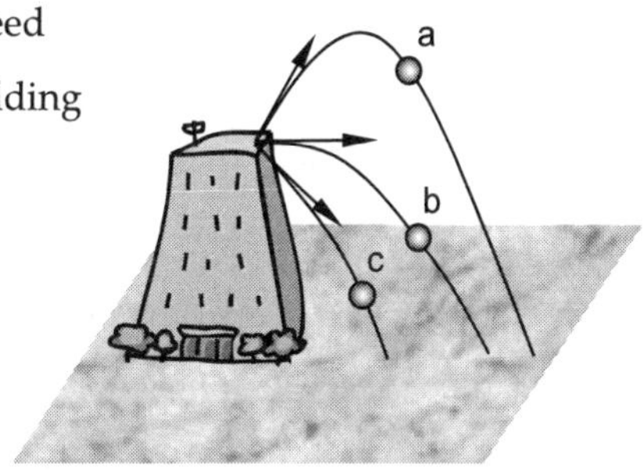

a) ball A
b) ball B
c) ball C
d) They all have the same speed at impact.

23. A crate is pulled up an inclined surface at constant speed by a motor as shown. If the crate had been pulled up at twice the original speed, then

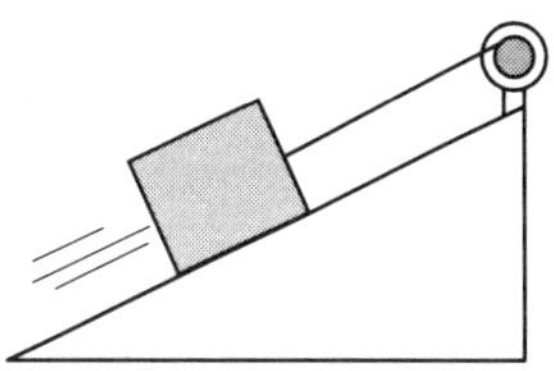

a) the crate's kinetic energy would have been doubled.
b) the work done by the motor would have been doubled.
c) the power exerted by the motor would have been doubled.
d) the efficiency of the motor would have been doubled.

Copyright © by June Chung - All rights reserved. Reproduction of this publication without the written permission of the copyright owner is strictly prohibited.

24. Which of the following units is not equivalent to one Watt [W]?

a) $1\ kg \cdot m^2/s^3$ b) $1\ N \cdot m/s$

c) $1\ J/s$ d) $1\ kg \cdot J^2/(N \cdot s)^2$

25 - 36. Written-response questions

25. An object at rest accelerates to 12 m/s when 5.8 J of work is done on it. If we want the same object to reach the speed of 36 m/s from rest, how much work must be done on the object?

26. A 2.0-kg object initially sliding on level ground at 10 m/s slows down and eventually comes to rest because of friction. How much work was done on the object by the friction force?

27. A 9.0-kg object is dropped from a height of 15 m above the ground. How high above the ground does the object's kinetic energy become twice its gravitational potential energy?

28. A 9.5-kg object is initially moving on level ground at 5.5 m/s, and we want to accelerate it to 8.2 m/s in 2.3 s. If the surface is frictionless, calculate (a) the magnitude of the horizontal force we need to apply on the object, (b) the work done on the object while we exert the applied force, and (c) the average power we need to exert on the object.

29. A 0.34-kg pendulum bob is released from the position as shown. What is the speed of the bob as the pendulum swings by the vertical position (indicated by the dotted line)? How will the answer change if the mass of the bob is doubled?

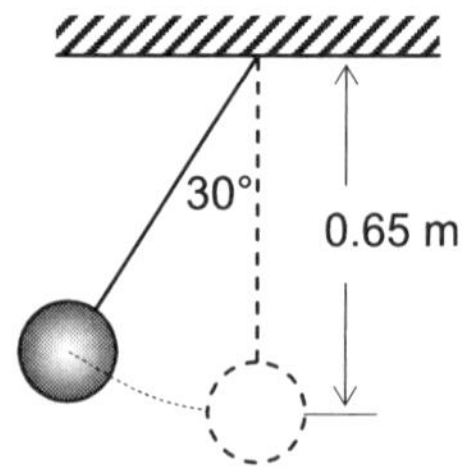

Copyright © by June Chung - All rights reserved. Reproduction of this publication without the written permission of the copyright owner is strictly prohibited.

30. A roller coaster car, initially moving at an unknown speed at point A, passes through point C at the speed of 27 m/s. What was the car's speed at point B? (Ignore friction.)

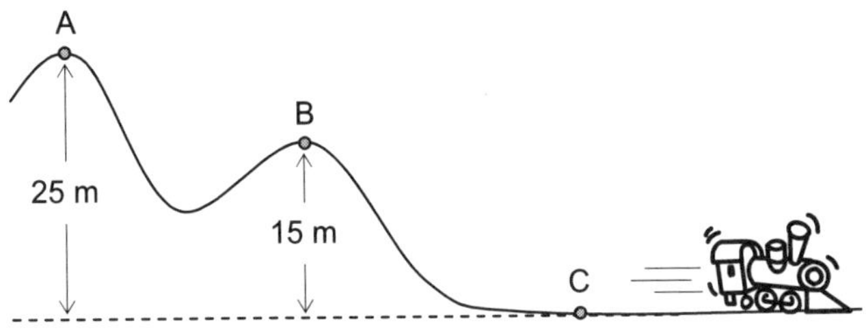

31. A 0.72-kg object is released from rest on an inclined surface as shown. If the object's speed at the bottom of the incline is 2.6 m/s, how much of its initial mechanical energy has been lost during the motion?

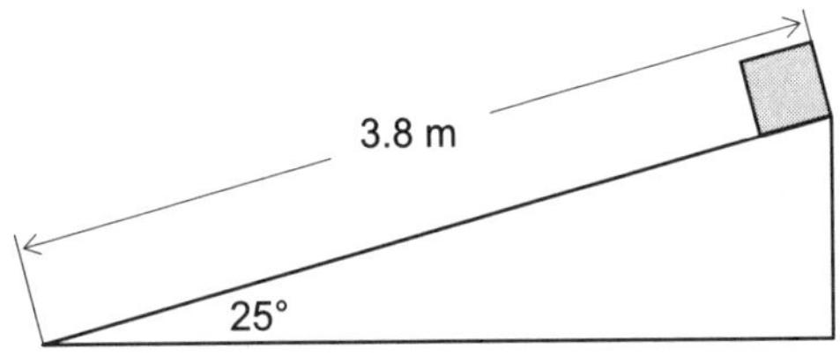

32. Two objects connected by a string are released from the position shown.

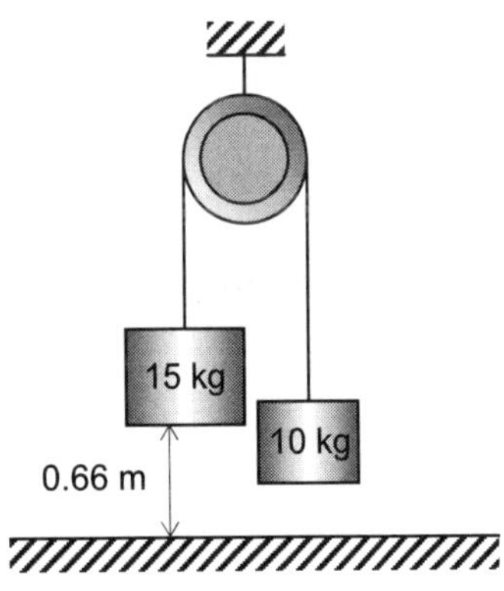

a) Find the change in each object's gravitational potential energy as the 15-kg object hits the ground, and add the changes to find the sum.

b) Multiply the sum you found in part (a) by -1 and let it equal the sum of the changes in the objects' kinetic energies. Then solve for the speed of the system as the 15-kg object hits the ground.

c) Applying the work-energy theorem to the 15-kg object, find the net work done on it.

d) Using the net work you found in part (c), find the magnitude of the tension force.

Copyright © by June Chung - All rights reserved. Reproduction of this publication without the written permission of the copyright owner is strictly prohibited.

33. An elevator driven by a 30-kW motor is carrying five occupants. The total mass of the elevator including the occupants is 1350 kg. If the elevator is pulled up at a constant speed of 0.85 m/s by the motor, what is the efficiency of the motor?

34. On average, the appliances in Phil's home are 65% efficient, and his monthly electricity consumption from using those appliances is 900 kW·h. If he upgrades his appliances and improves their average efficiency to 82%, how much will he save per month? (Assume the power company charges six cents per kilowatt-hour.)

35. A 1.5-kg object traveling at 1.2 m/s, north, collides with a 3.5-kg object moving at 0.50 m/s, south, and heads back south at 0.52 m/s after the collision. Determine whether the collision is perfectly elastic. If not, calculate how much of the objects' initial kinetic energy was lost during the collision.

36. A 7.4-kg object is moving east at 6.7 m/s when it collides with a 3.7-kg object at rest. If the collision is perfectly elastic, find the speed of each object after the collision.

37. A 2.5-kg block of wood is released from the top of an inclined surface as shown. When the block reaches the bottom of the incline, a 23-g bullet is fired horizontally at the block. The impact with the bullet causes the block to stop instantly. If the bullet's speed at the impact was 314 m/s, what is the height of the incline, *H*? (Assume the surface is frictionless.)

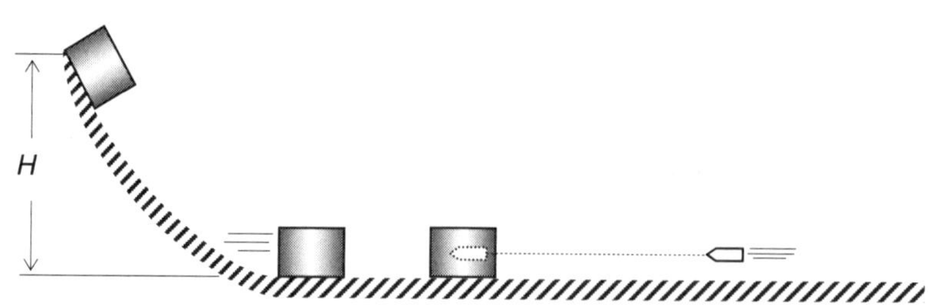

Copyright © by June Chung – All rights reserved. Reproduction of this publication without the written permission of the copyright owner is strictly prohibited.

CHAPTER 6

THERMAL PHYSICS

In this chapter it is important to understand the difference between thermal energy and heat, which are both energies and measured in Joules; and temperature, which states how hot a body is, and is measured in degrees Celsius or Kelvin. According to the kinetic molecular theory, all matter is made of atoms or molecules which are constantly in motion. These particles can move in a variety of ways, which gives them kinetic energy; and they are attracted and repulsed by each other, leading to potential energy. Although calculating the total amount of energy stored in a body is difficult, taking the temperature of a body is a simple matter, merely involving a thermometer.

Experience in the laboratory tells us that the temperature remains constant when a pure substance is changing state, even though heat may be pumped into or taken out of it. Also, we realize that it is easy to raise the temperature of some materials whilst others take longer time and more energy before they got hot. This leads us to consider what is happening on a molecular level as heat enters or leaves a material, and introduces the ideas of specific heat capacities and latent heats.

In this chapter, you will study:

- ✓ temperature, thermal energy, and heat, and how they are different
- ✓ specific heat capacity and latent heat
- ✓ conservation of heat

Copyright © by June Chung - All rights reserved. Reproduction of this publication without the written permission of the copyright owner is strictly prohibited.

Key Point 6.1: Thermal Energy (E_T), Temperature (T), and Heat (Q)

thermal energy, heat

When positive work is done on an object, its energy level increases, and it gains speed as a result (see Key Point 5.3, p. 93). However, **thermal energy** or heat delivered to a body seems to cause no change in its speed. (After all, you cannot make a pan move by heating it up.) In this case, it is the molecules of the body that gain speed. Thus, if you add the increase in kinetic energy of each molecule within the body, the sum will equal the amount of heat fed into it. Heat is often used interchangeably with thermal energy, but, to be exact, **heat** refers to the flow of thermal energy.

temperature

Heat into or out of an object causes a change in its **temperature**, which is proportional to the average kinetic energy of the molecules in the object. The greater the average molecular kinetic energy, the hotter the object. Temperature is a scalar quantity, and the SI unit for it is the Kelvin [K]. It is also measured in °C, which is related to K by

$$T\,(\text{K}) = T\,(^\circ\text{C}) + 273.15 \qquad \ldots (6.1)$$

thermal energy vs. temperature

Consider a cup of water and a bucket of water, both at 40 °C. While the average molecular kinetic energy of each is equal, the amount of thermal energy is greater in the latter simply because there are more water molecules in the bucket than in the cup. Hence, for the same material at a given temperature, the larger the mass, the greater the amount of thermal energy within the body. If two objects of the same material are at unequal temperatures, it is possible that the one with the smaller mass has a greater amount of thermal energy.

Now suppose we deliver an equal amount of heat into the cup and into the bucket. The temperature of the water in the cup will rise more because the thermal energy is distributed among fewer molecules (and therefore each molecule receives more).

> - Thermal energy within a body is the sum of all the molecular kinetic and potential energies.
> - Heat refers to the flow of thermal energy.
> - The temperature of a body is proportional to the average molecular kinetic energy.

Copyright © by June Chung – All rights reserved. Reproduction of this publication without the written permission of the copyright owner is strictly prohibited.

Key Point 6.2: Specific Heat Capacity (c)

quantifying thermal energy into or out of a body

The amount of heat (Q) required to cause a change in temperature (ΔT) of a given material is expressed as

$$Q = mc\Delta T \quad [\text{J} = \text{kg}\cdot(\text{J/kg C}^\circ)\cdot\text{C}^\circ] \quad \dots (6.2)$$

where c is the **specific heat capacity** of the material. According to Eq. (6.2), the value of Q is positive when the object's temperature rises (that is, when ΔT is positive) and negative when its temperature falls (that is, when ΔT is negative). Therefore, a positive value of Q indicates heat entering the object, and a negative value of Q means heat leaving the object.

effect of heat and mass on change in temperature

The equation also shows that the change in temperature is proportional to the amount of heat flowing into or out of the object. Thus, if the amount of heat into an object is doubled, the rise in temperature will double as well. In addition, the larger the mass, the more heat needs to be put into the object to realize a certain rise in temperature. For instance, the amount of heat required to raise the temperature of 2 kg of iron by 1 C° is twice the amount needed to raise the temperature of 1 kg of iron by 1 C°.

effect of specific heat capacity on change in temperature

However, if you put the same amount of heat into 1 kg of ice and 1 kg of iron, you will notice that the change in temperature is greater for iron than for ice. The reason is that iron has a lower specific heat capacity, which is the amount of thermal energy required to raise the temperature of one kilogram of the substance by 1 C°. Thus, it takes a lot of thermal energy to heat up or cool down a material with a large specific heat capacity. The specific heat capacities for various materials are listed in Table 6-1. Note that the SI unit for specific heat capacity is J/kg C°.

Table 6-1: Specific heat capacities of various substances

	Lead	Iron	Glass	Ice	Water	Steam
c [J/kg C°]	130	450	840	2100	4186	2010

Copyright © by June Chung – All rights reserved. Reproduction of this publication without the written permission of the copyright owner is strictly prohibited.

> - The amount of heat needed to change the temperature of a given material is expressed as
>
> $$Q = m c \Delta T$$
>
> $Q > 0$ ⟹ Heat flows <u>into</u> the object.
>
> $Q < 0$ ⟹ Heat flows <u>out of</u> the object.
>
> - The specific heat capacity, c, of a given substance is the amount of thermal energy needed to raise the temperature of 1 kg of the substance by 1 Celsius degree.

Example 6-1

A 3.5-kg block of hot lead, initially at 105 °C, is dropped into a bucket of ice water. After an hour, the temperature of the lead is measured to be 17 °C. Calculate the amount of heat lost by the lead.

Example 6-2

Suzan wants to find the specific heat capacity of an unknown solid, whose mass is 0.616 kg. She tries adding 8000 J of thermal energy to the solid and notices that its temperature rises from 20.5 °C to 36.0 °C. What is the specific heat capacity of the solid?

Copyright © by June Chung – All rights reserved. Reproduction of this publication without the written permission of the copyright owner is strictly prohibited.

Key Point 6.3: **Latent Heat (L)**

latent heat

Equation (6.2) in Key Point 6.2 is valid only if no change of phase occurs (say, from solid to liquid or from liquid to gas). During a change of phase, the temperature of a material stays constant even when heat is added to or drawn from the material, as shown in Fig. 6-1. When changing its phase, a material absorbs heat or gives up heat to alter the chemical bonds between the molecules. The amount of heat required to change the phase of 1 kg of a substance is called its **latent heat** and is measured in J/kg.

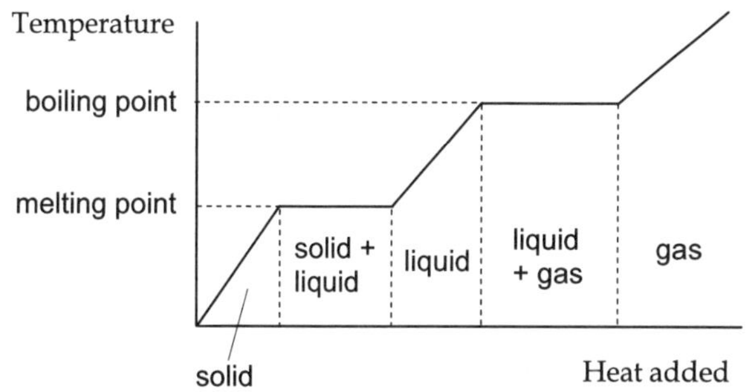

Figure 6-1: Temperature vs. heat added to a material

heat of fusion and heat of vaporization

There are two kinds of latent heat: **heat of fusion** and **heat of vaporization**. The former is the amount of heat required to change 1 kg of a substance from solid to liquid, and the latter is the amount of heat required to change 1 kg of a substance from liquid to gas. The latent heats for various substances are listed in Table 6-2.

Table 6-2: Latent heats for various substances

	Lead	Iron	Oxygen	Water
Melting point [°C]	327	1808	-219	0
Heat of fusion, L_f [J/kg]	0.25×10^5	2.89×10^5	0.14×10^5	3.33×10^5
Boiling point [°C]	1750	3023	-183	100
Heat of vaporization, L_v [J/kg]	8.7×10^5	63.4×10^5	2.1×10^5	22.6×10^5

example

To calculate the amount of heat required to turn 3.0 kg of ice initially at -10 °C into water at 25 °C, we need to add (1) thermal energy required to heat the ice to its melting point (Q_1), (2) thermal energy needed to melt

Copyright © by June Chung - All rights reserved. Reproduction of this publication without the written permission of the copyright owner is strictly prohibited.

the ice (Q_2), and (3) thermal energy required to heat the resulting water to 25 °C (Q_3):

$$\begin{aligned} Q_{total} &= Q_1 + Q_2 + Q_3 \\ &= (3.0)(2100)(0-(-10)) + (3.0)(3.33\times10^5) + (3.0)(4186)(25-0) \\ &= 1.38\times10^6 \text{ J} \end{aligned}$$

Thus, we need to add 1.38×10^6 J of thermal energy to turn 3.0 kg of ice at -10 °C into water at 25 °C. We will need to draw the same amount of energy out of 3.0 kg of water at 25 °C in order to turn it into ice at -10 °C.

> - Latent heat is the amount of heat required to change the phase of 1 kg of a substance.
> - Change of phase between solid and liquid ⟹ Use heat of fusion.
> - Change of phase between liquid and gas ⟹ Use heat of vaporization.

Example 6-3

If you want to change the temperature of a 4.75-kg block of ice from -12.0 °C to 12.0 °C, how much thermal energy do you need to deliver to the block of ice?

Example 6-4

0.52 kg of steam is initially at 110 °C. If 1.25×10^6 J of thermal energy is drawn out of it, what is the resulting phase and temperature of the matter?

Copyright © by June Chung – All rights reserved. Reproduction of this publication without the written permission of the copyright owner is strictly prohibited.

Key Point 6.4: Conservation of Heat

conservation of heat

As explained in Key Point 6.1, heat is thermal energy in motion. Heat can flow from one material to another in physical contact, but it can neither be created nor destroyed. Thus, the total amount of thermal energy within a system is conserved if there is no heat transfer between the system and its surroundings. In such cases, the heat lost by one part of the system will be gained by the other part. If a system consists of two parts A and B, the conservation of heat can be expressed as

$$Q_{lost\ by\ A} = Q_{gained\ by\ B} \qquad \ldots (6.3)$$

defining the system for conservation of heat

Remember that heat flows from the part at the higher temperature to the one at the lower temperature and continues to flow until the two reach an equilibrium temperature. For example, suppose we define a ball of hot iron and the water that surrounds it as the system. Then, being hotter than water, the iron will lose heat, all of which is absorbed by the water (assuming no heat enters or leaves the system). As a result, the iron's temperature drops as the water's temperature rises. However, if we had defined the ball of iron as our system, the total amount of thermal energy within that system would not be conserved because heat does escape into the surroundings (which is the water surrounding the iron). Thus, when applying **conservation of heat**, you need to define the system carefully so that there is no heat exchange across the boundary of the system.

> ➢ Conservation of heat:
> "The heat lost by one part of a system will be gained by another part if no heat is allowed to enter or leave the system."
>
> $$Q_{lost\ by\ A} = Q_{gained\ by\ B}$$

Copyright © by June Chung – All rights reserved. Reproduction of this publication without the written permission of the copyright owner is strictly prohibited.

Key Point 6.5: How to Solve Thermal Energy Problems

Thermal energy problems are relatively straightforward, so if you obtain the wrong answer, a mistake in your arithmetic is the most likely cause. Before you start solving a problem, rely on your experience and estimate what the final answer might be. Then, when you obtain the answer, compare it with your guess.

Approach thermal energy problems as outlined below:

1. Write down all the information given to you to determine the scope of the problem. If only one material is involved, go to Step 2a. Otherwise, go to Step 2b.

2a. Is there a change of phase? If not, use Eq. (6.2) to solve for the unknown. If there is, construct a table showing the amount of heat into or out of the material in each phase and during the change of phase. Watch out for the sign of the latent heat. It is negative if heat is released by the material. The sum is the total amount of heat involved in the entire process. An example is shown in Table 6-3 for 1.5 kg of steam at 120 °C condensing to water at 75 °C.

Table 6-3: Steam at 120 °C losing heat to become water at 75 °C

	ΔT	Heat (Q)
gas	-20 °C	$mc\,\Delta T = (1.5)(2010)(-20) = -60$ kJ
gas to liquid	0 °C	$mL = -(1.5)(22.6\times10^6) = -3390$ kJ
liquid	-25 °C	$mc\,\Delta T = (1.5)(4186)(-25) = -157$ kJ
Sum		-3600 kJ

2b. Define your system as the group of materials exchanging heat. Usually, there will be two materials, one giving up heat and the other receiving it. In Physics 11, we only consider problems where there is no change of phase for these materials. Then Eq. (6.3) can be written as

$$Q_{lost\ by\ A} = -m_A c_A (T_f - T_{iA}) = m_B c_B (T_f - T_{iB}) = Q_{gained\ by\ B} \quad \ldots (6.4)$$

Note that the final temperature is the equilibrium temperature, reached by both materials. Also, the negative sign is needed on the

Copyright © by June Chung - All rights reserved. Reproduction of this publication without the written permission of the copyright owner is strictly prohibited.

left-hand side to balance the right-hand side ($Q_{lost\ by\ A}$ is negative and $Q_{gained\ by\ B}$ is positive). Set up Eq. (6.4), make sure you have only one unknown in the equation, and solve for it.

3. Draw a graph similar to Fig. 6-1 to represent the events taking place in the system, and ask yourself if your answer makes sense. Compare it with your initial estimate.

Example 6-5

A block of hot iron is dropped into a bucket of water. After a few minutes, the drop in temperature of the iron is measured to be 8.1 C° while the rise in temperature for the water is 1.4 C°. Assuming all the heat lost by the iron was absorbed by the water, determine the ratio of the mass of the iron to that of the water.

Example 6-6

An unknown quantity of thermal energy, when added to 0.95 kg of water at 20 °C, turns it into steam at 100 °C. If this quantity of heat is added to 0.95 kg of iron at 20 °C, will it be enough to melt at least part of the iron? Show your calculations.

Copyright © by June Chung – All rights reserved. Reproduction of this publication without the written permission of the copyright owner is strictly prohibited.

Example 6-1	Use Eq. (6.2) to calculate the amount of heat given up by the lead: $$Q = (3.50)(130)(17-105) = -40040 \text{ J}$$ Hence, the lead lost 40 kJ of thermal energy. (The negative sign indicates that heat came out of the lead.)
Example 6-2	Use Eq. (6.2) to find the specific heat capacity of the block: $$8000 = (0.616)c(36.0-20.5);\; c = 839 \text{ J/kg·C°}$$
Example 6-3	The amount of heat you need to add to ice in order to raise its temperature from -12 °C to 0 °C: $$Q_1 = (4.75)(2100)(0-(-12)) = 119700 \text{ J}$$ The amount of heat you need to add to ice in order to change its phase from solid to liquid: $$Q_2 = (4.75)(3.33\times10^5) = 1581000 \text{ J}$$ The amount of heat you need to add to ice in order to raise its temperature from 0 °C to 12 °C: $$Q_3 = (4.75)(4186)(12-0) = 238600 \text{ J}$$ The total amount of thermal energy you need to deliver to ice is thus $$Q_1 + Q_2 + Q_3 = 1.94\times10^6 \text{ J}$$
Example 6-4	First, find the amount of heat needed to be drawn out of the steam to cool it down to the boiling point of 100 °C (Q_1): $$Q_1 = (0.52)(2100)(110-100) = 10920 \text{ J}$$ Q_1 is about 0.87% of 1.25×10^6 J. Next, find the amount of heat needed to be drawn out of the steam to turn it into water (Q_2) using heat of vaporization: $$Q_2 = (0.52)(22.6\times10^5) = 1.13\times10^6 \text{ J}$$ We still need to draw 0.11×10^6 J of energy from the water. Inspect whether this is sufficient to cool its temperature to 0 °C: $$Q_3 = (0.52)(4186)(100-0) = 0.22\times10^6 \text{ J}$$ Since Q_3 is greater than 0.11×10^6 J, we cannot cool the water to 0 °C. The final temperature of the water can be obtained from $$0.11\times10^6 = (0.52)(4186)(100-T)$$ The final temperature of the water is 50 °C.

Copyright © by June Chung - All rights reserved. Reproduction of this publication without the written permission of the copyright owner is strictly prohibited.

Example 6-5

The water and iron are exchanging heat, so define the system to include both materials. Then the amount of heat lost by the iron, $Q_{out\,of\,Iron}$, is equal to that received by the water, $Q_{to\,Water}$. We can set up Eq. (6.4) as follows:

$$-m_{iron}(450)(-8.1) = m_{water}(4186)(1.4)$$

Solve for $\frac{m_{iron}}{m_{water}}$, and we find that the ratio is 1.6 to 1.

Example 6-6

Find the amount of heat added to turn the water into steam. Since a change of phase occurs, make a table as shown:

	ΔT	Heat (Q)
liquid	80 °C	$mc\Delta T = (0.95)(4186)(80) = 318$ kJ
liquid to gas	0 °C	$mL = -(0.95)(22.6\times10^6) = 2147$ kJ
gas	0 °C	0 kJ
Sum		2465 kJ

When 2465 kJ of heat is added to the iron at 20 °C, its temperature rises according to Eq. (6.2):

$$2465000 = (0.95)(450)(T_f - 20);\ T_f = 5790\ ^\circ\text{C}$$

Because the final temperature is higher than the melting point of iron, the iron melts. (Note: Our analysis so far does not show whether the iron melts completely, but we can say that it melts at least partially.)

Copyright © by June Chung - All rights reserved. Reproduction of this publication without the written permission of the copyright owner is strictly prohibited.

PROBLEMS ON CHAPTER 6: THERMAL PHYSICS

(Answer Key on p. 232)

1 – 5. Fill in the blanks

1. __________ is a measure of coldness or hotness and is proportional to the average molecular kinetic energy.

2. ________ ________ is defined as the sum of individual molecular kinetic and potential energies.

3. ________ ________ ________ of a material is the amount of thermal energy required to raise the temperature of 1 kg of the material by 1 C°.

4. ________ ________ is the amount of heat required per 1 kg of mass to cause a change of phase.

5. Heat of __________ is associated with the change of phase between liquid and gas.

6 – 11. True/False questions

6. _____ The SI unit for temperature is the degree Celsius.

7. _____ Substances with less thermal energy are colder than those with more thermal energy.

8. _____ The greater the specific heat capacity of a substance, the easier it is to raise its temperature.

9. _____ Heat of vaporization is greater than heat of fusion.

10. _____ During a change of phase, a material does not absorb or give up thermal energy because there is no change in temperature.

11. _____ The conservation of heat principle states that two objects in contact with each other will eventually reach the same temperature.

12 – 16. Multiple choice questions

12. What property makes water a suitable substance for coolant?

    a) Water has a higher density than most liquids.
    b) Water boils at a higher temperature than most liquids.
    c) Water requires a lot of thermal energy to raise its temperature.
    d) Water is abundant.

Copyright © by June Chung – All rights reserved. Reproduction of this publication without the written permission of the copyright owner is strictly prohibited.

13. Two different substances with identical mass are initially at different temperatures. We can compare their specific heat capacities by measuring

    a) how much heat each substance absorbs from or gives up to its surroundings.
    b) how fast each substance's temperature changes initially.
    c) the change in each substance's temperature while they are in contact.
    d) the equilibrium temperature of each substance.

14. Two different solids, initially at 20 °C, are heated to 50 °C. If both solids had the same mass and there was no phase change for either solid, which of the following statements is true?

    a) The solid with the higher specific heat received more thermal energy.
    b) The solid with the lower specific heat received more thermal energy.
    c) They both received the same amount of thermal energy.
    d) We cannot determine which solid received more thermal energy.

15. Three solids at different temperatures were brought into contact. Solid A was initially at 10 °C, solid B at 40 °C, and solid C at 90 °C. After a while, an equilibrium temperature of 50 °C was reached, with no heat exchange between the solids and their surroundings. Which of the following statements is false?

    a) The amount of heat given up by solid C was absorbed by solids A and B.
    b) Solid A received the most thermal energy.
    c) The total amount of thermal energy within the solids was conserved.
    d) Solid B received thermal energy.

16. If the temperature of a substance does not change even when it receives 150 kJ of thermal energy, which of the following is true?

    a) The substance has a very low specific heat capacity.
    b) The substance's heat of fusion is 150 kJ.
    c) The substance is in the gas phase.
    d) The substance is either at its melting point or boiling point.

17 – 25. Written-response questions

17. How much thermal energy is required to turn 350 g of ice at 0.00 °C to steam at 100 °C?

Copyright © by June Chung – All rights reserved. Reproduction of this publication without the written permission of the copyright owner is strictly prohibited.

18. Water is supplied to a boiler that can deliver 8700 kJ of thermal energy per hour to the water. If the boiler must raise the water's temperature by at least 30 C°, what is the maximum mass of water the boiler can heat in one hour?

19. An insulated container contains 5.7 kg of water at 60 °C. If you want to cool the temperature of the water to 45 °C by adding water at 12 °C, how much water should you add?

20. A 7.5 kg of solid iron, initially at 1500 °C, is immersed in a pool containing 21 kg of water at 10 °C. Find the equilibrium temperature if no heat enters or leaves the system.

21. 0.222 kg of an unknown substance at 111 °C is dropped into 0.345 kg of water at 35.7 °C. If the equilibrium temperature is 46.8 °C, what is the specific heat capacity of the substance?

22. 2.3 kg of an unknown liquid at 8.5 °C is mixed with 0.55 kg of the same liquid at 49 °C. What is the final temperature of the mixture?

23. A 1000-W espresso machine can heat 1.0 kg of water from 20 °C to 50 °C in 3.0 minutes. Find the efficiency of the machine.

24. A 275-g cube of ice, initially at -15 °C, is dropped into a glass container at 65 °C. If the equilibrium temperature of the system is 15 °C, what is the mass of the glass container? (Be careful: There is a change of phase for the water.)

25. A block of iron is dropped from 1.80 m above the surface of water as shown. Once immersed in the water, the iron sinks towards the bottom of the container at the speed of 0.23 m/s. Assuming no heat enters or leaves the system, find the change in temperature of the system when the iron is about to touch the bottom of the container. (m_{iron} = 1.7 kg, m_{water} = 3.5 kg)

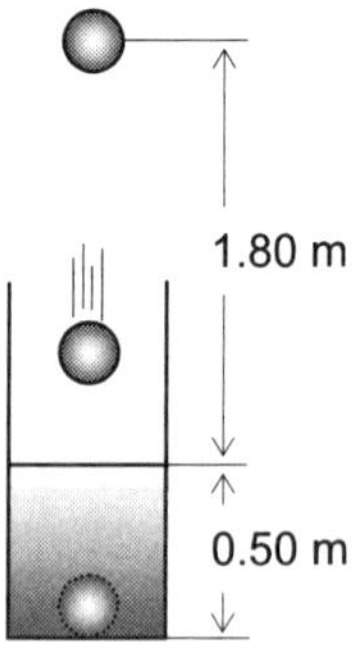

Copyright © by June Chung – All rights reserved. Reproduction of this publication without the written permission of the copyright owner is strictly prohibited.

CHAPTER 7

WAVE

We know that energy is a conserved quantity and that objects can transfer energy in various ways (for example, by colliding with each other). Then we can use conservation of energy and momentum to calculate the objects' energy and velocity after the collision. However, this is not the only way to transfer energy. A wave is another way of transporting energy, usually through a material substance called a medium, but not always. You can create waves by hitting water with a spoon or your finger, and the energy of your blow is transmitted through the water by means of a wave. Waves have some unique properties that enable us to understand how electrons occupy the space within an atom, or how we can simultaneously hear and pick out the sounds made by violins, trumpets, and drums when listening to music.

In this chapter, you will study:

✓ the nature of the wave

✓ transverse waves and longitudinal waves

✓ amplitude, wavelength, frequency, and period of a wave

✓ the wave equation

✓ properties of waves: reflection, refraction, diffraction, and interference

✓ total internal reflection

✓ standing waves

✓ the Doppler effect

Copyright © by June Chung - All rights reserved. Reproduction of this publication without the written permission of the copyright owner is strictly prohibited.

Key Point 7.1: Waves

waves

A **wave** is a disturbance that carries energy from one place to another. If a wave needs to travel through a **medium** like air or water, it is called a **mechanical wave**. Examples are sound waves, water waves, and earthquakes. If a wave can travel through a vacuum, it is called an **electromagnetic wave**.† Light, radio waves, microwaves, and X-rays are electromagnetic waves.

source and propagation of waves

All waves are created by some kind of vibration, such as a vibrating tuning fork or vocal cord. For mechanical waves, energy must be delivered to one part of the medium to create a disturbance. The particles that receive the energy transfer it to their neighbors, which do the same to their neighbors, and so on. In this way, energy can move from one part of the medium to another. Note that the particles themselves do not move along the path of the energy. They merely transmit the energy.

example

To understand this concept better, suppose a pebble is dropped from above the surface of a pond. As the pebble falls, its gravitational potential energy is converted into kinetic energy. When the pebble hits the surface of the water, its energy is used to create circular ripples, which then spread outward. Thus, the energy given by the pebble to the water molecules is traveling outward in the form of ripples. However, the water molecules do not travel with the ripples. They rise up when a ripple is about to pass them and move down when it just passes them.

- A wave is a disturbance that carries energy from one location to another.
- Mechanical waves travel through a medium; Electromagnetic waves can travel through a vacuum.
- A wave transports energy without transporting matter. That is, energy travels as a wave, but particles in the medium do not travel in the direction of the wave.

† Electromagnetic waves are created through interaction between the electric and magnetic fields produced by oscillating charged particles, rather than interaction between particles in a medium. Since a full understanding of this phenomenon requires a higher level of math and physics, a detailed description is not given here.

Copyright © by June Chung – All rights reserved. Reproduction of this publication without the written permission of the copyright owner is strictly prohibited.

Key Point 7.2: Transverse Wave vs. Longitudinal Wave

For mechanical waves, the particles in a medium can vibrate in one of two ways: perpendicular to the direction of energy travel or parallel to it. If the particles' vibration is perpendicular to the wave propagation, the wave is called a **transverse wave**. For example, you can create a transverse wave by repeatedly jerking a horizontal rope up and down, and the shape of the resulting wave is shown in Fig. 7-1. The wave form in the figure is typical of a transverse wave. The equilibrium position, shown with the dashed line, is the position of the rope when it is not disturbed. Points A and B, which have the maximum upward displacement from the equilibrium position, are called **crests**, and C and D, which have the maximum downward displacement, are called **troughs**. The distance from the equilibrium position to a crest (or a trough) is defined as the **amplitude**. The amplitude of a wave is related to the energy being carried by that wave. (For example, a loud sound wave has a large amplitude whereas the amplitude of a soft sound wave is small.) The length of one complete wave cycle is called the **wavelength**, measured in meters per cycle [m/cycle]. It is usually easiest to find the wavelength by measuring the distance between two successive crests (or two successive troughs).

transverse waves

anatomy of transverse waves

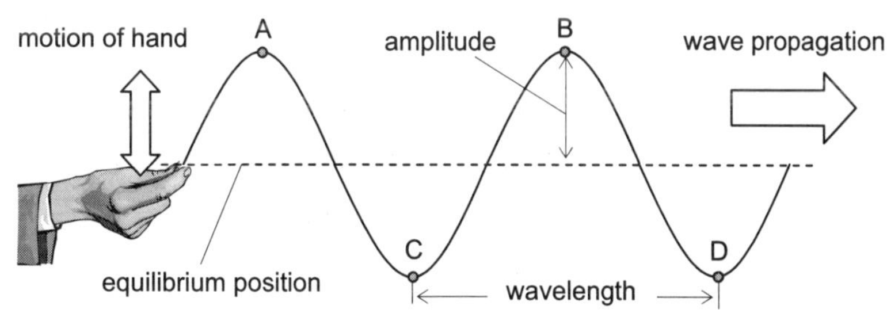

Figure 7-1: A transverse wave generated by an up-and-down hand motion

longitudinal waves

If the particles of the medium vibrate back and forth, parallel to the direction of energy travel, the wave is called a **longitudinal wave**. A sound wave is a longitudinal wave as the air molecules vibrate back and forth along the direction of the wave propagation. This movement of the air particles creates variations in density, as shown in Fig. 7-2.

Copyright © by June Chung – All rights reserved. Reproduction of this publication without the written permission of the copyright owner is strictly prohibited.

anatomy of longitudinal waves

Regions with the maximum and minimum density are called **compressions** and **rarefactions**, respectively. The maximum displacement of the particles from their equilibrium positions is the amplitude of the wave, and the distance between two successive compressions (or rarefactions) is the wavelength. The alternating pattern of compressions and rarefactions is one of the characteristics of longitudinal waves.

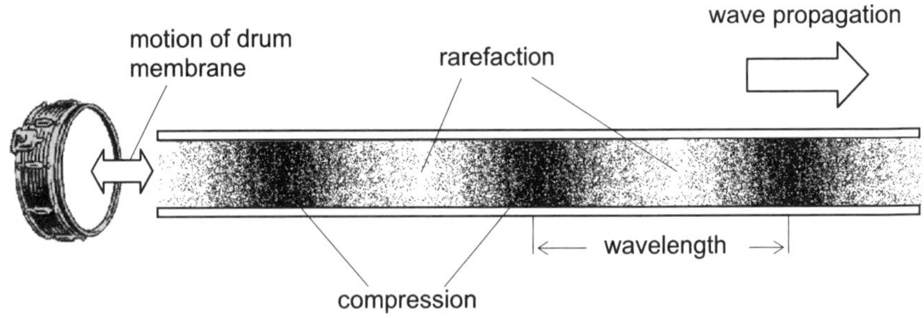

Figure 7-2: A longitudinal wave created by the back-and-forth motion of a drum membrane

- In a transverse wave, particles of the medium vibrate perpendicular to the wave propagation.
- In a longitudinal wave, particles vibrate back and forth parallel to the wave propagation.
- A transverse wave has an alternating pattern of crests and troughs; a longitudinal wave has an alternating pattern of compressions and rarefactions.
- Amplitude is the maximum displacement of a particle from its equilibrium position.
- Wavelength is the length of one complete wave cycle.

Copyright © by June Chung – All rights reserved. Reproduction of this publication without the written permission of the copyright owner is strictly prohibited.

Key Point 7.3: Frequency (f) and Period (T) of a Wave

creation of transverse waves

When one end of a horizontal rope is moved up and down periodically, a transverse wave is created and travels away from that end of the rope, as shown in Fig. 7-3. It may seem as if points on the rope also move in the direction of the wave, but they can only move up or down (see how points a, b, c, and d move in the figure).

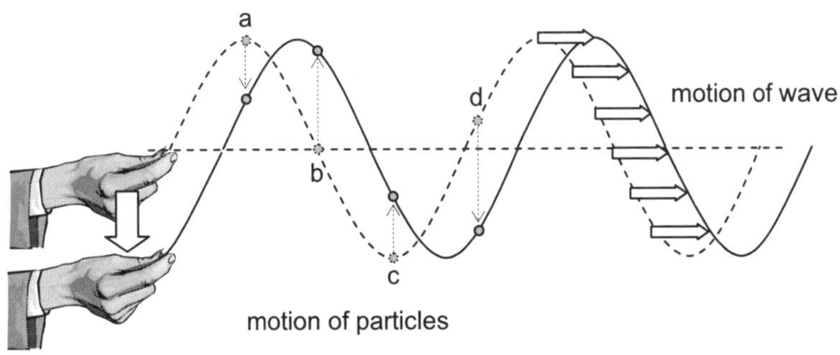

Figure 7-3: A traveling wave

frequency

If you stand at the other end of this rope, you will see the wave coming toward you. You can measure the **frequency** of this wave by counting how many wave cycles reach you in one second. Alternatively, you can measure the frequency by counting how often the hand (which is the vibrating energy source) vibrates in one second because the frequency of a wave is always the same as the frequency of its source. The frequency is measured in Hertz [Hz], which is equivalent to cycles/second, or simply s^{-1}.

period

The reciprocal of frequency is **period**, the time required for one complete wave cycle, and it is measured in seconds. If the wave in the rope has a low frequency, it will take a long time for one cycle to pass you. Hence, the higher the frequency of a wave, the shorter its period.

> - Frequency is the number of cycles per second [Hz = cycles/s].
> - Period is the time for one complete cycle [s/cycle].
> - Frequency (f) is the reciprocal of period (T), and vice versa:
>
> $$f = \frac{1}{T} \; ; \; T = \frac{1}{f}$$

Copyright © by June Chung - All rights reserved. Reproduction of this publication without the written permission of the copyright owner is strictly prohibited.

Example 7-1

Determine the frequency and the period of rotation of (a) a second hand and (b) a minute hand on a clock.

Example 7-2

A wave has an amplitude of 0.25 m, and the distance from a crest to the nearest trough is 0.12 m. The frequency of the wave is 16 Hz. Determine the wavelength and the period of the wave.

Copyright © by June Chung – All rights reserved. Reproduction of this publication without the written permission of the copyright owner is strictly prohibited.

Key Point 7.4: Wave Equation

universal wave equation

The speed of a wave can be determined by measuring the distance it travels in a given time and dividing it by that time. With the wavelength and frequency of a wave, we can find its speed from

$$v = \lambda f = \frac{\lambda}{T} \text{ [m/s]} \quad \ldots (7.1)$$

where λ is the wavelength and f is the frequency. This equation is known as the **universal wave equation**, and it applies to all waves (both mechanical and electromagnetic).

example

For example, if a wave has a frequency of 2 Hz, it means that two complete cycles pass a particular point in one second. If its wavelength is 3 m, the length of two complete cycles is 6 m. Thus, we can see that the wave travels six meters in one second, and its speed is 6 m/s.

relationship between frequency and wavelength

At a given temperature, the speed of a wave through a particular medium is constant. For instance, the speed of sound in air at room temperature is about 340 m/s. Sound propagates much faster in helium gas, traveling at about 1000 m/s. The speed of sound through water is even faster at about 1440 m/s. Thus, it is important to realize that, since all sound waves travel at the same speed through a particular medium, the waves with higher frequencies have shorter wavelengths, and those with lower frequencies have longer wavelengths. This applies to electromagnetic waves, which travel at a constant speed of 3.00×10^{8} m/s in a vacuum. The ones with higher frequencies such as UV-rays and X-rays have relatively short wavelengths, but the ones with lower frequencies such as radio waves and microwaves have much longer wavelengths.

> - Universal wave equation:
>
> $$v = \lambda f = \frac{\lambda}{T}$$
>
> - v is determined by the medium; f is determined by the energy source. λ links these two fixed values together according to the above equation.

Copyright © by June Chung – All rights reserved. Reproduction of this publication without the written permission of the copyright owner is strictly prohibited.

Example 7-3

Find the wavelength of a sound wave with the frequency of 277 Hz. The speed of sound is 343 m/s (assume the medium is air at room temperature unless stated otherwise).

Example 7-4

The frequency of the microwaves used in wireless phones is about 1.00 GHz. Find the wavelength. The speed of electromagnetic waves is 3.00×10^{8} m/s.

Copyright © by June Chung – All rights reserved. Reproduction of this publication without the written permission of the copyright owner is strictly prohibited.

Key Point 7.5: **Reflection**

When a wave reaches a barrier through which it cannot go, some of its energy is transferred to the barrier, but most of it is reflected back into the medium (for example, sound waves hitting a wall). To show the reflection of a wave graphically, first draw the **wave fronts** of a wave by connecting the crests, as shown in Fig. 7-4. The direction of wave travel is always perpendicular to the wave fronts and can be indicated with lines called **rays**. In Fig. 7-4, the wave fronts are parallel to the x-axis while the wave propagates along the y-axis.

wave fronts and rays

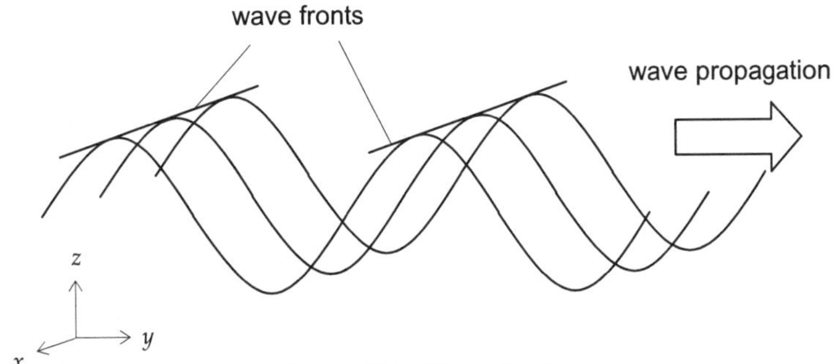

Figure 7-4: Wave fronts

When viewed from directly above the xy-plane, the wave can be represented as a set of wave fronts and a ray, as shown in Fig. 7-5. In the figure, the **incident ray** is shown along with the **angle of incidence**. The angle is always measured between the ray and a line perpendicular to the boundary drawn at the point where the ray meets the barrier. (This perpendicular line is called the **normal** and is just a construction

description of reflection in Fig. 7-5

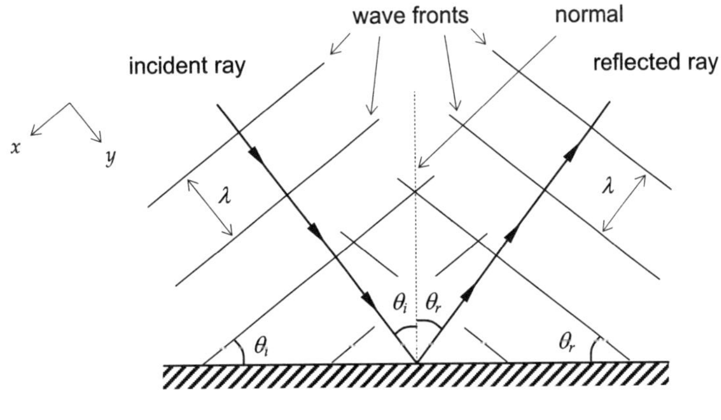

Figure 7-5: Reflection of a wave

Copyright © by June Chung – All rights reserved. Reproduction of this publication without the written permission of the copyright owner is strictly prohibited.

device that allows us to measure the angles of incidence and reflection easily.) Thus, if the angle of incidence is zero, it means that the wave fronts are parallel to the surface of the boundary. Knowing the orientation of the incident ray, you can then draw the **reflected ray** so that the **angle of reflection** is equal to the angle of incidence, as can be seen in

law of reflection

Fig. 7-5. This is the **law of reflection**, which can be written as

$$\theta_i = \theta_r \qquad \ldots (7.2)$$

Note that, although the direction of travel changes as a result of reflection, the speed, wavelength, and frequency of the wave all remain constant. The wave's amplitude decreases if some of its energy is transferred to the barrier.

> - When a wave reaches a barrier, most of its energy is reflected, but some is transferred to the barrier.
> - The law of reflection:
>   "Waves always reflect off a barrier so that the angle of incidence is equal to the angle of reflection."

Copyright © by June Chung - All rights reserved. Reproduction of this publication without the written permission of the copyright owner is strictly prohibited.

Key Point 7.6: **Refraction**

refraction

When a wave traveling in one medium crosses a boundary into another medium with a different density, its speed changes. The denser the medium, the slower the speed of the wave. Hence, the medium with the higher density is called the slow medium, and the one with the lower density is the fast medium. The speed of a wave thus increases when entering a faster medium and decreases when entering a slower one. Such a change in speed causes the wave to **refract**, or change its direction when entering a second medium at an angle.

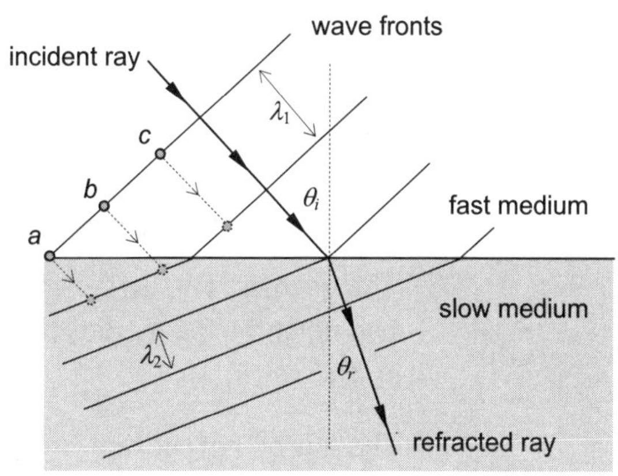

Figure 7-6: Refraction of a wave

description of refraction in Fig. 7-6

Figure 7-6 shows a wave refracting at the boundary when entering a slower medium. Points *a*, *b*, and *c* are initially on a straight wave front approaching the boundary. Since point *a* is already at the boundary, it now travels at a reduced speed. After a few moments, point *b* crosses the boundary and slows down while point *c* has not yet experienced any change in speed. Because different parts of the wave front experience a change in speed at different times, the wave front bends at the boundary as shown. Note that the wavelength of the wave is reduced in the slower medium. Since the frequency of the wave cannot change once the wave is set in motion, the wavelength must decrease if the speed is reduced, in accordance with the universal wave equation ($\lambda_2 < \lambda_1$ in the figure).

Copyright © by June Chung – All rights reserved. Reproduction of this publication without the written permission of the copyright owner is strictly prohibited.

law of refraction

The amount of bending, or refraction, depends on the difference in speeds between the two media. The greater the difference in speed, the greater the amount of refraction. Exactly how much a wave refracts can be obtained from the **law of refraction**:

$$\frac{\sin\theta_i}{\sin\theta_r} = \frac{v_i}{v_r} \quad \ldots (7.3)$$

According to Eq. (7.3), if a wave slows down upon entering a new medium, the ray always refracts towards the normal. On the other hand, when a wave enters a faster medium, the ray always refracts away from the normal.

- A wave slows down upon entering a denser medium.
- When crossing a boundary at an angle, a wave refracts because parts of the wave front experience the change in speed at different times.
- The greater the difference in density between two neighboring media, the greater the refraction of the wave.
- The law of refraction: $\frac{\sin\theta_i}{\sin\theta_r} = \frac{v_i}{v_r}$
- When entering a faster medium at an angle, a ray refracts away from the normal (i.e., $\theta_i < \theta_r$), and wavelength and speed increase.
- When entering a slower medium at an angle, a ray refracts towards the normal (i.e., $\theta_i > \theta_r$), and wavelength and speed decrease.

Example 7-5

A wave crosses a boundary between two different liquids, and its speed increases from 25 m/s to 33 m/s. If the angle of incidence is 14°, what is the angle of refraction?

Copyright © by June Chung - All rights reserved. Reproduction of this publication without the written permission of the copyright owner is strictly prohibited.

Key Point 7.7: Total Internal Reflection

refraction of waves entering a faster medium

While a wave can always enter a slow medium from a faster one, it cannot always pass from a slow medium into a fast one. Recall that rays refract away from the normal when they enter a faster medium, which means that the **angle of refraction** is greater than the angle of incidence. Therefore, as the angle of incidence increases, the angle of refraction must increase as well in order to satisfy the law of refraction in Eq. (7.3).

critical angle

When the angle of incidence is large enough, the angle of refraction becomes 90°, and the refracted ray would skim along the boundary, as illustrated in Fig. 7-7. The angle of incidence at which this occurs is called the **critical angle**, θ_c. For any angle of incidence greater than the critical angle, the ray does not refract at all. Instead, it is reflected back internally, following the law of reflection in Eq. (7.2). This phenomenon is thus called **total internal reflection**.

total internal reflection

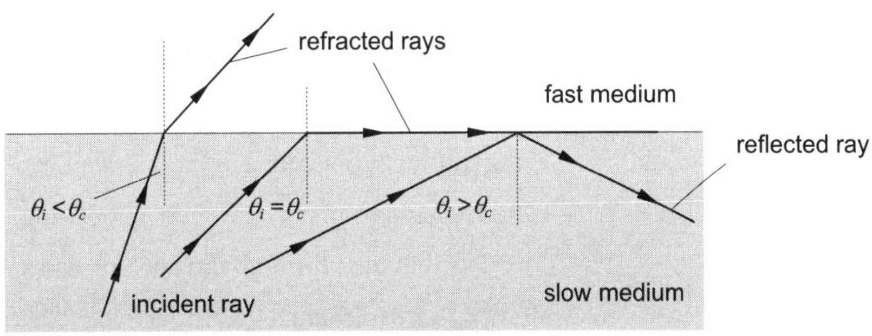

Figure 7-7: Refraction of a wave from a slow medium into a fast medium

> - When a wave enters a faster medium, the ray always refracts away from the normal.
> - Critical angle = the angle of incidence at which the angle of refraction becomes 90 degrees
> - For incidence angles greater than the critical angle, all rays are reflected internally.

Example 7-6

A light wave, upon entering a new medium, speeds up by 50% of its initial speed. Calculate the critical angle.

Copyright © by June Chung – All rights reserved. Reproduction of this publication without the written permission of the copyright owner is strictly prohibited.

Key Point 7.8: Diffraction

diffraction

Rays usually travel in straight lines. However, when they pass through an opening or come to a corner, they change the direction of travel as shown in Fig. 7-8. This change in direction of rays is called **diffraction**. (Note that the wavelength, speed, and frequency all remain constant because the medium does not change.)

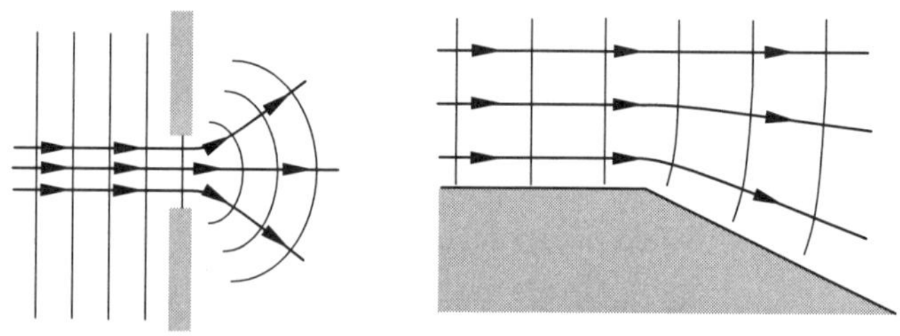

Figure 7-8: Diffraction of a wave

Huygen's principle

One way to explain how diffraction occurs is to imagine a wave front as a series of little wavelets, as shown in Fig. 7-9. When by itself, each wavelet tends to spread outward, but when aligned next to each other along a straight line, none of the wavelets can spread. When a wave front encounters an obstacle with an opening, as in Fig. 7-9, those wavelets that pass through the opening can spread in all directions, allowing rays to change their directions. This explanation is known as **Huygens' principle**.

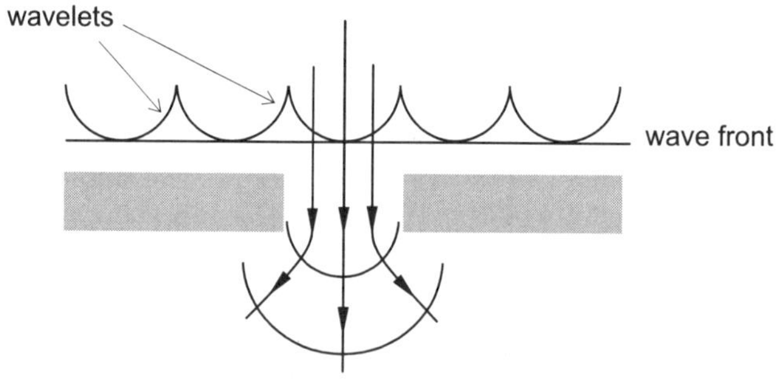

Figure 7-9: Wave front as a series of wavelets

effect of wavelength and size of opening on diffraction

The amount of diffraction depends on the ratio of the wavelength of the wave to the size of the opening or the obstacle. The higher the ratio, the greater the diffraction. For this reason, sound waves can diffract quite

Copyright © by June Chung – All rights reserved. Reproduction of this publication without the written permission of the copyright owner is strictly prohibited.

significantly while visible light (whose wavelength is in the order of a few hundred nanometers) hardly diffracts at all (unless the opening is extremely small).

> - Diffraction occurs when a wave passes through an opening or by an obstacle.
> - Diffraction can be explained with Huygens' principle, which models a wave front as a series of little wavelets.
> - The amount of diffraction is proportional to wavelength and inversely proportional to the size of the opening or the obstacle.

Example 7-7

Two sound waves whose frequencies are 350 Hz and 700 Hz, respectively, pass through a small opening. Compare the amount of diffraction that occurs.

Copyright © by June Chung - All rights reserved. Reproduction of this publication without the written permission of the copyright owner is strictly prohibited.

Key Point 7.9: Interference

interference

Two or more waves or pulses can pass through the same medium at the same time without affecting each other. Only the particles of the medium are momentarily affected. This is called **interference** and is a special property of waves. When waves interfere with one another, the new wave pattern that is produced can be predicted using the **principle of superposition**, which states that the resultant displacement of a particle is determined by adding the displacements of the individual waves.

principle of superposition

constructive interference

There are two kinds of interference. The first is called **constructive interference**. This happens when the interfering waves have a displacement in the same direction, and an example is shown in Fig. 7-10. The resultant displacement from constructive interference is always greater than the displacement of each wave.

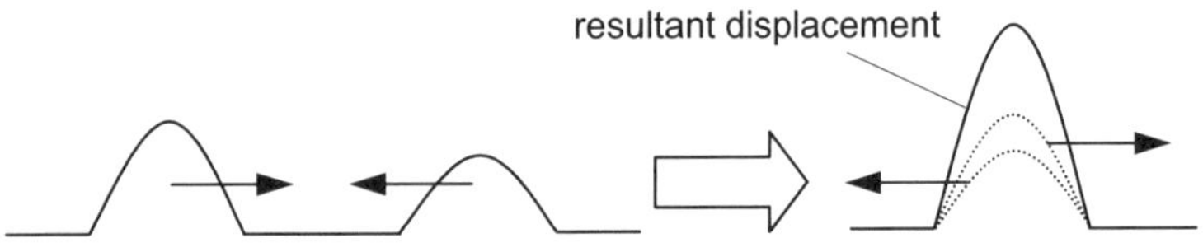

Figure 7-10: Constructive interference of two wave pulses

destructive interference

The second is called **destructive interference**, and this happens when the interfering waves have a displacement in opposite directions. With destructive interference, the resultant displacement is the difference in the displacements of the waves, as shown in Fig. 7-11. Thus, if the two waves had an identical displacement in opposite directions, the resultant displacement would be zero.

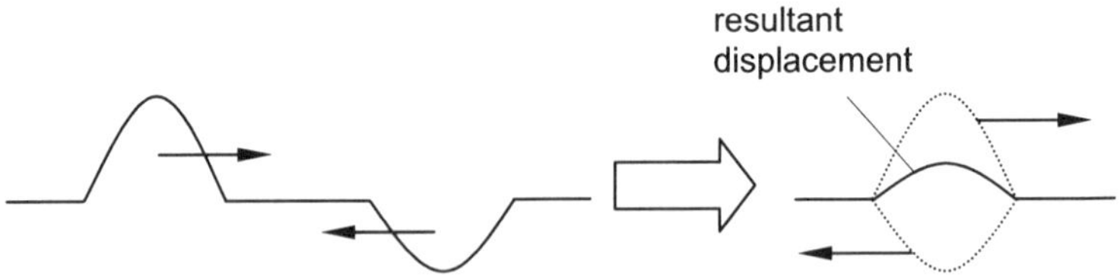

Figure 7-11: Destructive interference of two wave pulses

Copyright © by June Chung - All rights reserved. Reproduction of this publication without the written permission of the copyright owner is strictly prohibited.

- Waves interfere with one another when they pass through the same medium simultaneously.
- The resultant displacement is determined by the principle of superposition.
- Constructive interference occurs when the interfering waves have a displacement in the same direction.
- Destructive interference occurs when the interfering waves have a displacement in opposite directions.

Example 7-8

Draw the resulting displacement when the two pulses overlap.

(The pulses have the same width.)

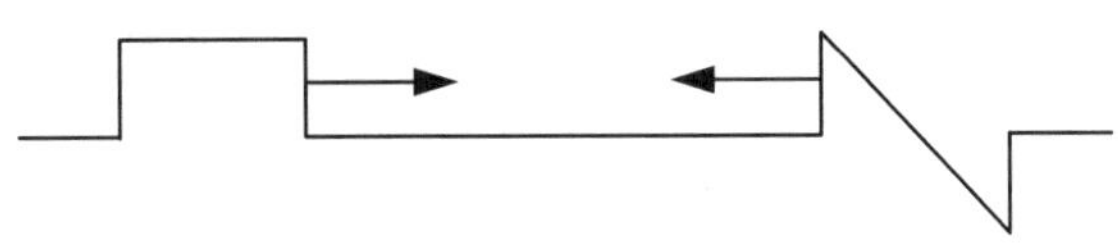

Copyright © by June Chung - All rights reserved. Reproduction of this publication without the written permission of the copyright owner is strictly prohibited.

Key Point 7.10: **Reflection at a Boundary and the Formation of Standing Waves**

inversion of a pulse at a boundary

When a wave crosses a boundary between two media having different densities, the energy carried by the wave is partially reflected (see Key Point 7.5). The shape of the reflected wave becomes inverted if the incident wave was traveling in the less dense medium. The inversion of a wave pulse is illustrated in Fig. 7-12. Note that the amplitude of the reflected pulse is less than that of the incident pulse because part of the energy carried by the incident pulse has been transmitted to the denser medium. If the difference in density between the media is fairly large, little energy will be transmitted to the denser medium, and the amplitude of the reflected pulse will be nearly the same as that of the incident pulse.

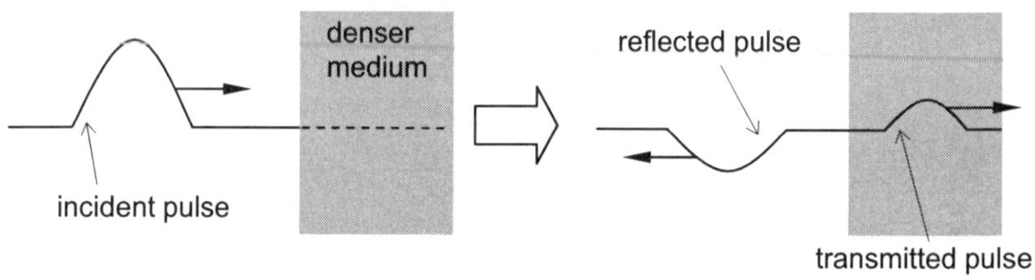

Figure 7-12: Inversion of reflected wave pulse

interference between incident and reflected pulses

If two wave pulses approach a denser medium, the first to be reflected will interfere with the second, which is yet to be reflected, as shown in Fig. 7-13. Now, imagine a wave traveling in one medium bounded by a denser medium as in Fig. 7-14. The wave will be reflected back and forth, and, as it does, continuous interference will occur. Such interference is complex as there will be a combination of constructive and destructive interference.

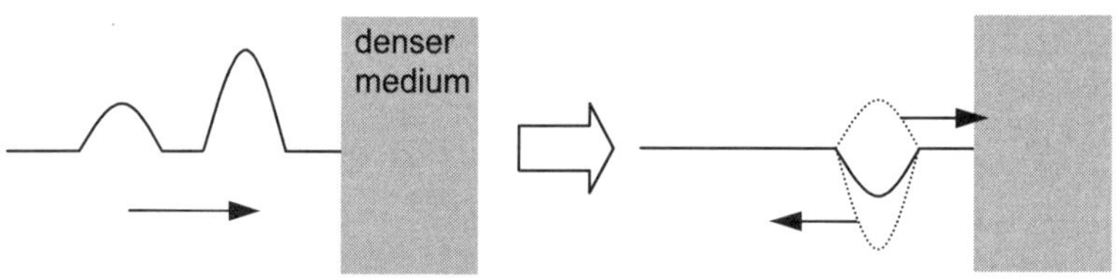

Figure 7-13: Interference between an incident pulse and a reflected pulse

Copyright © by June Chung – All rights reserved. Reproduction of this publication without the written permission of the copyright owner is strictly prohibited.

standing wave pattern

If the incident and reflected parts of the wave have the same wavelength and amplitude, the continuous interference between them creates **standing wave** patterns at certain frequencies known as **natural frequencies**, as shown in Fig. 7-14. Although it appears that the wave is standing still in each pattern, it is not an actual wave. It is just a wave-shaped pattern, with nodes and anti-nodes. A **node** is a point where the particles of the medium remain at rest because the interfering waves always undergo destructive interference there. An **anti-node** (or **loop**) is a point where the particles suffer the maximum displacement because the interfering waves always undergo constructive interference there.

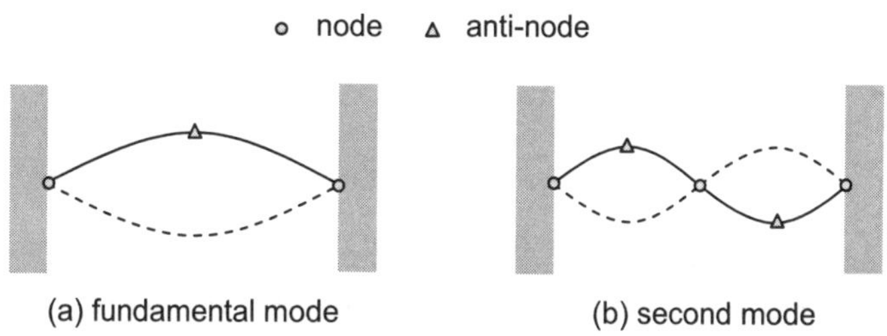

Figure 7-14: Standing wave patterns

mode and frequency of a standing wave

The lowest frequency at which a standing wave pattern is formed is called the **fundamental frequency**, and the shape of the pattern at that frequency is called the **fundamental mode**. As shown in Fig. 7-14 (a), this pattern contains only half a wavelength. The second mode is shown in Fig. 7-14 (b), and it contains a whole wavelength. Since the wave speed is constant within a medium, this means that the frequency of the second mode is twice the fundamental frequency. Similarly, the frequency of the third mode is three times the fundamental frequency, and so on. Hence, we find the following relationship between the frequency associated with the n-th mode and the fundamental frequency:

$$f_n = n f_1 \qquad \dots (7.4)$$

where f_1 is the fundamental frequency and f_n is the frequency of the n-th mode.

Copyright © by June Chung - All rights reserved. Reproduction of this publication without the written permission of the copyright owner is strictly prohibited.

- A reflected wave becomes inverted if the incident wave was traveling in the less dense medium.
- When a wave is reflected at a boundary, the reflected part interferes with the incident part.
- At natural frequencies, such interference creates a standing wave pattern.
- The lowest natural frequency is called the fundamental frequency; this pattern, which contains one-half the wavelength, is called the fundamental mode.
- The frequency of the *n*-th mode is related to the fundamental frequency by $f_n = n f_1$.

Example 7-9

After studying the first two modes in Fig. 7-14, draw the third mode. Given that the width of the medium carrying the wave is 2.4 m and the wave speed is 340 m/s, find the frequencies of the first three modes.

Example 7-10

A standing wave pattern with four nodes is observed in a 9.0-m long string that is fixed at both ends. If the wave speed is 340 m/s, find the frequency of the standing wave.

Copyright © by June Chung – All rights reserved. Reproduction of this publication without the written permission of the copyright owner is strictly prohibited.

Key Point 7.11: **Doppler Effect**

wave created by a stationary source

When the source of a wave is stationary, the wave fronts form concentric circles† with the source at their center, as can be seen in Fig. 7-15 (a). In this case, since the wavelength, which is the distance between two consecutive wave fronts, is the same in all directions, the frequency of the wave would be the same no matter where the receiver is located with respect to the source.

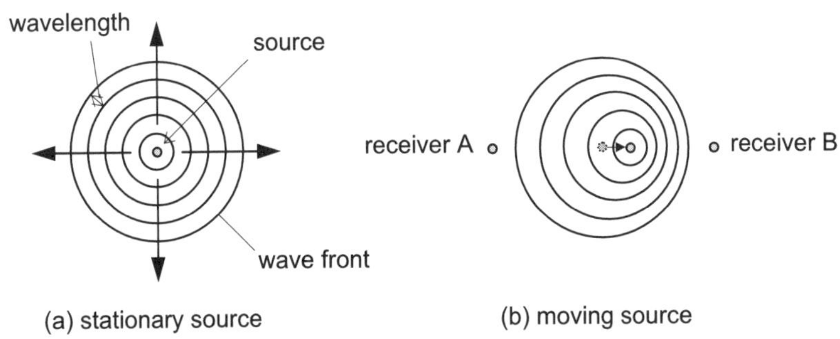

Figure 7-15: Doppler effect for a moving source

Doppler effect due to source's motion

However, for a moving source, the wave fronts no longer form concentric circles, as shown in Fig. 7-15 (b). That is, the wave fronts ahead of the source are closer together while those behind it are farther apart than they were for a stationary source. This means that, a receiver ahead of a moving source detects more wave cycles passing by it in a given time while a receiver behind it detects fewer cycles in the same time. Therefore, the frequencies of the wave received by A and B in Fig. 7-15 (b) will be lower and higher, respectively, than the actual frequency. Note that the source is emitting a wave whose frequency stays constant whether it is moving or not. It is the receiver who perceives a change in frequency because of the source's motion.

Doppler effect due to receiver's motion

The same effect can be achieved when a receiver is moving toward or away from a stationary source. This is the **Doppler effect**, which states that "When either the source or the receiver moves relative to the other, the receiver perceives a change in the frequency of the wave."

† The wave fronts form concentric spheres for three-dimensional waves.

Copyright © by June Chung – All rights reserved. Reproduction of this publication without the written permission of the copyright owner is strictly prohibited.

Namely, when the source or the receiver moves away from the other, the receiver perceives a drop in frequency. When they get closer, the receiver perceives an increase in frequency.

application of Doppler effect

The Doppler effect has been used to prove that our universe is expanding and the rate of expansion is increasing. Basically, scientists observed the electromagnetic waves emitted by distant stars near the boundary of our universe. If the universe was collapsing, we, the receivers of the waves, would notice an increase in frequency because the boundary would be moving closer to us. If the universe was expanding at a constant rate, we would notice a drop in frequency, but that drop in frequency would not change with time. However, scientists found that the drop in frequency was increasing with time, which indicated that the expansion of our universe is accelerating.

Doppler effect in ambulance siren

The Doppler effect can also be used to explain why you perceive a drop in the frequency of a siren when an ambulance passes. While the ambulance is approaching, you hear a higher frequency than the actual one. When it passes you by, the pitch of the siren suddenly drops because it is now moving away from you.

> - Doppler effect:
>   "When either the source or the receiver moves relative to the other, the receiver detects a change in frequency of the wave."
> - When the source or the receiver moves away from the other, the receiver perceives a drop in frequency; when they get closer, the receiver perceives an increase in frequency.

Example 7-11

A source is emitting a sound wave with frequency of 300 Hz and is initially 200 m to the west of the receiver. Describe the frequency perceived by the receiver if it is moving east at 35 m/s while the source is moving east at 20 m/s.

Copyright © by June Chung – All rights reserved. Reproduction of this publication without the written permission of the copyright owner is strictly prohibited.

ANSWER KEY TO EXAMPLES

Example 7-1	(a) It takes 60 seconds for a second hand to complete one cycle. Hence, its period is 60 seconds, and its frequency is 0.0167 Hz. (b) It takes 60 minutes, or 3600 seconds for a minute hand to complete one cycle. Hence, its period is 3600 seconds, and its frequency is 0.000278 Hz.
Example 7-2	The distance of 0.12 m is exactly half the length of one complete cycle. The wavelength of the wave is therefore 0.24 m. The period of the wave is simply the reciprocal of the frequency, which is 16 Hz. Thus, the period is 0.063 s.
Example 7-3	Use the wave equation: $343 = \lambda(277)$ The wavelength is 1.24 m.
Example 7-4	Use the wave equation: $3.00\times10^{8} = \lambda\left(1.00\times10^{9}\right)$ The wavelength is 0.300 m.
Example 7-5	Use the law of refraction: $\frac{\sin 14°}{\sin\theta_r} = \frac{25}{33}$ Solve for $\sin\theta_r$, and we find it is 0.319. To calculate the angle of refraction, use the inverse sine: $\theta_r = \sin^{-1}(0.319) = 19°$
Example 7-6	At the critical angle, the angle of refraction is 90°. The ratio of velocities on the right-hand side of Eq. (7.3) is 0.67. Thus, $\frac{\sin\theta_C}{\sin 90°} = 0.67$ Since $\sin 90° = 1$, $\sin\theta_C = 0.67$. To calculate the critical angle, use the inverse sine: $\theta_C = \sin^{-1}(0.67) = 42°$

Copyright © by June Chung – All rights reserved. Reproduction of this publication without the written permission of the copyright owner is strictly prohibited.

Example 7-7	According to the wave equation in Eq. (7.1), an increase in frequency causes the wavelength to decrease. If the frequency of the sound wave is doubled, the wavelength is reduced in half. The ratio of the wavelength to the size of the opening, therefore, is also halved, and the amount of diffraction decreases.
Example 7-8	Draw the pulses together and add the displacement of each pulse algebraically at various points. The resulting shape is drawn below: 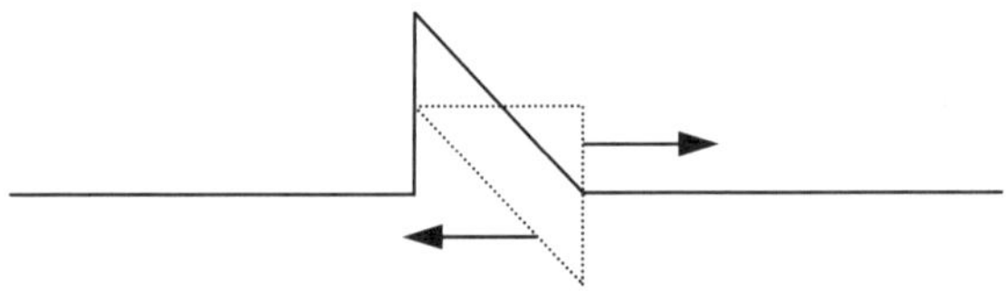
Example 7-9	The third mode is shown at right: 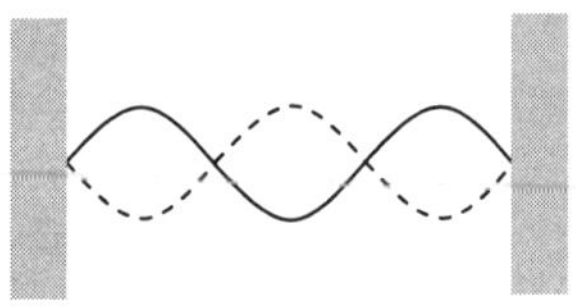 Since the fundamental mode contains a half wavelength, the wavelength of the fundamental mode is 4.8 m. Using the wave equation, we can find the fundamental frequency: $$f_1 = \frac{v}{\lambda} = \frac{340}{4.8} = 71 \text{ Hz}$$ The frequencies of the second and third modes are obtained by applying Eq. (7.4): $$f_2 = 2f_1 = 142 \text{ Hz}$$ $$f_3 = 3f_1 = 213 \text{ Hz}$$
Example 7-10	The standing wave pattern developed in the string is the third mode because there are four nodes. As shown in the figure above in Example 7-9, the wavelength of the third mode is two-thirds of the string's length, or 6.0 m. The frequency of the standing wave can be obtained by using the wave equation: $$f_3 = \frac{v}{\lambda} = \frac{340}{6.0} = 57 \text{ Hz}$$
Example 7-11	Relative to the source, the receiver is moving away from it at 15 m/s. Hence, the receiver perceives a lower frequency.

Copyright © by June Chung - All rights reserved. Reproduction of this publication without the written permission of the copyright owner is strictly prohibited.

PROBLEMS ON CHAPTER 7: WAVE

(Answer Key on p. 233)

1 - 12. Fill in the blanks

1. A _________ is a disturbance that propagates from one location to another.
2. A _________ wave requires a medium to propagate.
3. In ________ waves, the particles' vibration is perpendicular to wave propagation.
4. The length of one complete wave cycle is called the _________.
5. The regions with maximum and minimum density in a longitudinal wave are called _________ and _________.
6. The time needed for one wave cycle to pass through a particular point is called the _________.
7. The speed of a wave can be obtained from the product of the _________ and the _________ of the wave.
8. The _________ of a wave occurs when the wave enters a new medium with a different density at an angle.
9. There are two kinds of interferences: _________ and _________ interferences.
10. A _________ in a standing wave is the point with no displacement.
11. The lowest frequency at which a standing wave pattern is formed is called the _________ frequency.
12. The _______ _______ occurs when the receiver perceives a change in frequency of the wave because of its relative movement with respect to the source.

13 - 25. True/False questions

13. _____ Infrared waves are electromagnetic waves.
14. _____ Sound waves are transverse waves.
15. _____ Frequency is the reciprocal of period.
16. _____ The speed of a wave through a particular medium is constant at a given temperature.
17. _____ When a wave is reflected off a barrier, the angle of reflection is always equal to the angle of incidence.

Copyright © by June Chung - All rights reserved. Reproduction of this publication without the written permission of the copyright owner is strictly prohibited.

18. _____ When entering a fast medium from a slow medium, a wave always refracts towards the normal.

19. _____ The greater the difference in density between two neighboring media, the greater the amount of refraction of a wave.

20. _____ Total internal reflection occurs when the angle of incidence is less than the critical angle.

21. _____ Diffraction occurs when two waves interfere with each other.

22. _____ Constructive interference occurs when the interfering waves have a displacement in the same direction.

23. _____ Standing waves are possible only for transverse waves.

24. _____ The second mode of a standing wave has twice the wavelength of the fundamental mode.

25. _____ When the source of a wave moves away from the receiver, the receiver detects a change in the frequency of the wave.

26 - 34. Multiple choice questions

26. Which of the following waves is <u>not</u> a mechanical wave?

a) Sound b) Light c) Ocean wave d) Earthquake

27. When a wave reaches the boundary of a medium, some of the energy is ________. Which of the following may be used to complete the above statement?

a) amplified temporarily
b) destroyed forever
c) bound at the boundary
d) reflected back into the medium

28. A straw is dipped into the water in a glass as shown. When you observe the straw from above the surface of the water, the straw looks as if

a) it is shorter than its normal length
b) it is longer than its normal length
c) its length has not changed
d) it is thinner than its normal thickness

Copyright © by June Chung – All rights reserved. Reproduction of this publication without the written permission of the copyright owner is strictly prohibited.

29. The answer in Problem 28 can be explained by

a) reflection of light
b) refraction of light
c) diffraction of light
d) interference of light waves

30. The answer in Problem 29 occurs because

a) the frequency of the wave changes at the surface of the water.
b) the amplitude of the wave changes at the surface of the water.
c) the speed of the wave changes at the surface of the water.
d) the period of the wave changes at the surface of the water.

31. We know sound waves diffract because

a) we cannot hear sound when the source of the wave is too far
b) we can hear sound around a corner
c) we cannot hear sound when two sound waves interfere destructively.
d) we detect a change in frequency when the source is moving relative to us.

32. Which of the following statements about interference is always true?

a) The resultant displacement is doubled at the point of constructive interference.
b) The resultant displacement is zero at the point of destructive interference.
c) The resultant displacement may be zero at the point of constructive interference.
d) The resultant displacement is determined by the principle of superposition.

33. Standing wave patterns result from

a) the reflection of a wave
b) the refraction of a wave
c) the interference between the incident part and the reflected part of a wave
d) the diffraction of a wave

34. The Doppler effect does <u>not</u> occur if

a) neither the source nor the receiver moves
b) the source moves while the receiver is stationary
c) the receiver moves while the source is stationary
d) both the source and the receiver move but at different speeds

Copyright © by June Chung – All rights reserved. Reproduction of this publication without the written permission of the copyright owner is strictly prohibited.

35 - 41. Written-response questions

35. A wave moving at the speed of 1500 m/s has a wavelength of 2.5 m. Find the frequency and the period of the wave.

36. In a longitudinal wave, the distance from a compression to the nearest rarefaction is 1.2 m. If five complete wave cycles pass through a point per second, find the speed of the wave.

37. A wave initially traveling at 340 m/s enters a slower medium at an angle of 35° from the normal. If the angle of refraction is 20°, what is the speed of the wave in the slower medium?

38. A sound wave traveling in air is approaching a water-air boundary. If the angle of incidence is 10°, will it be able to penetrate into water? Why or why not? Explain. (The speed of sound through the air is 340 m/s; the speed of sound through water is 1440 m/s.)

39. Draw the resulting displacement when the two pulses overlap. (The pulses have the same width.)

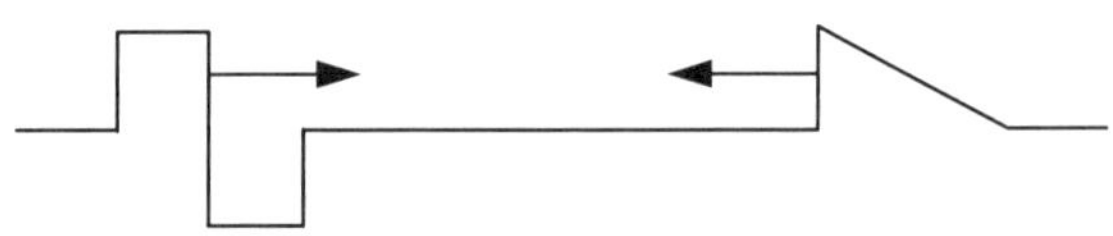

40. A standing wave has three anti-nodes and a wavelength of 0.24 m. If the wave speed is 340 m/s, find the wavelengths and the frequencies of the first two modes.

41. (a) Two objects of mass 50 g each are moving directly towards each other with speeds of 10 m/s and 6.0 m/s, respectively. Describe and calculate the results when and after they meet. (Assume a perfectly inelastic collision.)

(b) Two energy pulses are moving toward each other through a medium. They both have a wavelength of 5.0 cm and an amplitude of +3.5 cm. Describe and show what happens when and after they meet.

Copyright © by June Chung - All rights reserved. Reproduction of this publication without the written permission of the copyright owner is strictly prohibited.

CHAPTER 8

GEOMETRIC OPTICS

Having investigated the properties of waves, we can now apply our knowledge to light. But what exactly is light? Is it a stream of tiny bundles of energy called photons, or is it a wave? The debate continues to this day as both theories have some truth in them. Wherever the truth lies, we know that light can be reflected and refracted, and this knowledge has many practical applications. Images are formed using mirrors and lenses, and applying the laws of reflection and refraction, we can understand how periscopes work, and what kind of lens we need in our glasses.

In this chapter, you will study:

- ✓ how an image is formed
- ✓ image formation by a plane mirror
- ✓ image formation by spherical mirrors
- ✓ mirror equation and lateral magnification
- ✓ image formation by thin lenses
- ✓ lens equation

Copyright © by June Chung – All rights reserved. Reproduction of this publication without the written permission of the copyright owner is strictly prohibited.

Key Point 8.1: Image Formation

ray model of light

Visible light, being an electromagnetic radiation, shows all the wave characteristics including diffraction. However, the amount of diffraction is negligible under most circumstances, and we may assume that light rays travel in straight lines. This is the **ray model of light**, which is adopted throughout this chapter.

requirements for us to see objects

This chapter deals with the image formation of objects using mirrors and lenses, but let us first understand how it is that we can see things. In order for our eyes to see an object, the following three requirements must be satisfied:

1. There must be a light source. The Sun, lamps, and fire are examples of light sources.
2. There must be an object that absorbs certain segments of the visible light spectrum (which is made up of the colors red, orange, yellow, green, blue, indigo, and violet) and reflects others. For instance, a ripe tomato is red because it absorbs all segments of the spectrum except the one associated with the color red, which it reflects. Objects like window glass do not absorb any part of the light spectrum, so we do not see them.
3. Our eyes must be in position to receive the light rays reflected from the object. Since the surface of most objects is not perfectly smooth, light rays bounce off them in all directions. (This is called **diffuse reflection**.) Thus, the light rays can reach our eyes as long as we are looking in the right direction.

> - Under most circumstances, light rays may be assumed to travel in straight lines.
> - For us to see an object, we need:
>   1. A light source to generate visible light.
>   2. An object that absorbs certain segments of the light spectrum and reflects others.
>   3. Our eyes in a position to receive the light rays bounced off the object.

Copyright © by June Chung – All rights reserved. Reproduction of this publication without the written permission of the copyright owner is strictly prohibited.

Key Point 8.2: Image Formation by a Plane Mirror

use of mirrors

One way to see objects is to turn our eyes towards them and receive the light rays reflected from them. Another way is to use mirrors. Using a plane mirror, we can see images of the objects around us without having to turn our head. We can instead turn the mirror so that images of the desired objects appear.

image formation by a plane mirror

Let's see how an image appears in a plane mirror. First of all, light rays emitted by a light source are reflected from an object and scatter in all directions, as illustrated in Fig. 8-1. Some of these rays are incident upon the plane mirror and are reflected according to the law of reflection. Our eyes can receive these reflected rays if they are in the proper position. If not, we cannot see the image of the object.

characteristics of a plane mirror image

Such an image appears to be on the other side of the mirror because our brain does not know that the rays reaching our eyes have been reflected. Instead, it gets tricked into thinking that the rays traveled in a straight path from behind the mirror. The result is an image formed behind the mirror with the same size as the object. Because the image is not really there and is formed by our brain's misinterpretation, it is called a **virtual image**. A **real image** is one through which light rays actually pass. For an image formed by a plane mirror, the **image distance** d_i is equal to the **object distance** d_o, as shown in Fig. 8-1. Therefore, the position of the image does not change even if we observe it from various places (as long as the reflected light can enter our eyes).

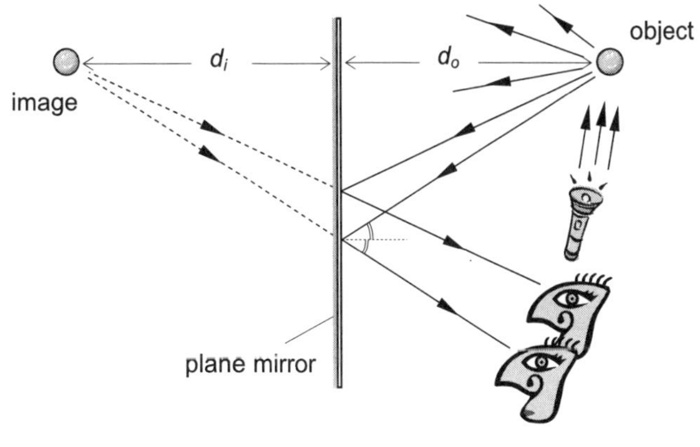

Figure 8-1: Image formation by a plane mirror

Copyright © by June Chung - All rights reserved. Reproduction of this publication without the written permission of the copyright owner is strictly prohibited.

An image formed by a plane mirror is...

- a virtual image
- as large as the object itself
- located as far behind the mirror as the object is in front of it

Example 8-1

Two plane mirrors joined at a right angle are used to see the images of a tennis ball, as shown below. By drawing a ray diagram, determine the locations of the virtual images (there are a total of three images).

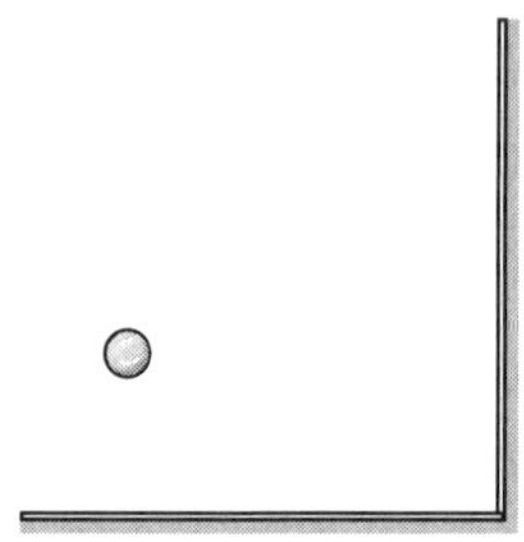

Copyright © by June Chung - All rights reserved. Reproduction of this publication without the written permission of the copyright owner is strictly prohibited.

Key Point 8.3: Spherical Mirrors

spherical mirrors

Curved mirrors whose surfaces form a section of a sphere are called **spherical mirrors**. A spherical mirror can be either concave or convex. Light rays reflected from a concave mirror converge to a real focus while those reflected from a convex mirror diverge from a virtual focus, as shown in Fig. 8-2.

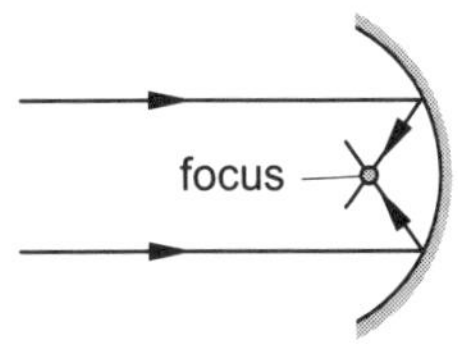

(a) concave mirror

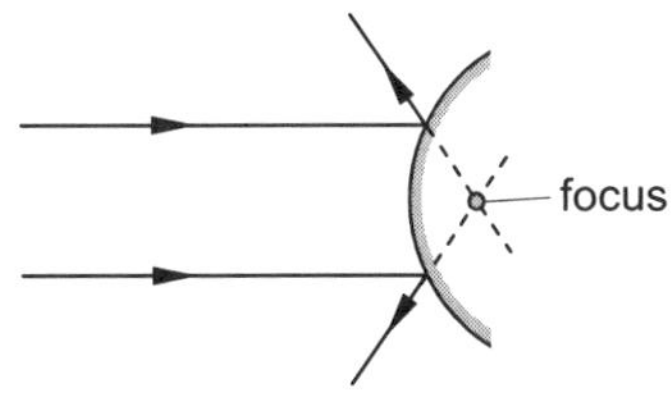

(b) convex mirror

Figure 8-2: Reflection of light rays by spherical mirrors

anatomy of spherical mirror

To understand how an image is formed by these spherical mirrors, you need to know the following terms (refer to Fig. 8-3).

1. The **center of curvature** (C) is the center of the sphere from which the mirror was cut. Thus, the distance from the mirror to the center of curvature is the radius of that sphere. For a concave mirror, the center of curvature is located in front of it. For a convex mirror, the center of curvature is located behind it.
2. The **vertex** (*V*) is the geometric center of the mirror.
3. Located halfway between the center of curvature (C) and the vertex (V) is the **focal point** (*F*), and the distance from the vertex to the focal point is known as the **focal length**. (Thus, $\overline{VF} = 0.5\overline{VC}$ in Fig. 8-3.)
4. The line passing through the vertex, the focal point, and the center of curvature is called the **principal axis**.

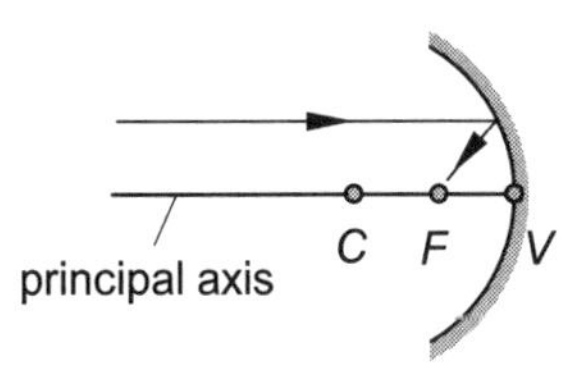

Figure 8-3: Anatomy of a concave mirror

Copyright © by June Chung - All rights reserved. Reproduction of this publication without the written permission of the copyright owner is strictly prohibited.

- Light rays reflected from a concave mirror converge; those reflected from a convex mirror diverge.
- Center of curvature = center of the sphere from which the mirror is cut
- Vertex = geometric center of the mirror
- Focal point = midpoint between the vertex and the center of curvature
- Principal axis = the line passing through the center of curvature, the vertex, and the focal point

Example 8-2

A convex mirror is cut from a sphere whose diameter is 0.88 m. First draw the principal axis of the mirror, and indicate the vertex, the focal point, and the center of curvature. Also, find the focal length of the mirror.

Copyright © by June Chung – All rights reserved. Reproduction of this publication without the written permission of the copyright owner is strictly prohibited.

Key Point 8.4: Image Formation by a Concave Mirror

rules for drawing a ray diagram for concave mirrors

The formation of an image by a **concave mirror** can be understood by drawing two rays according to these rules (refer to Fig. 8-4):

1. Draw an incident ray parallel to the principal axis from the object to the mirror. Upon reflection, this ray passes through the focal point.[†]
2. Draw an incident ray connecting the object, the focal point, and the mirror. Upon reflection, this ray travels parallel to the principal axis.

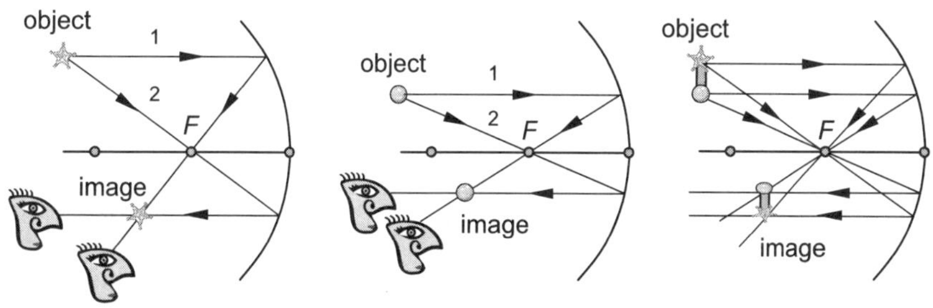

Figure 8-4: Image formation by a concave mirror

The image is formed where these two reflected rays meet. (In fact, all reflected rays from the object, including those not drawn in Fig. 8-4, pass through the point where the image is formed.) Hence, the location of the image will be the same to all observers no matter where they are. With the first two diagrams in Fig. 8-4, we can determine the image of any long object, as shown in the last diagram.

characteristics of a concave mirror image

Using the method described above, we can always find the image of an object regardless of its location. For a concave mirror, the characteristics of the image depend on how far the object is from the mirror. For instance, if the object is located beyond the center of curvature (*C*), as in Fig. 8-4, an inverted, real image appears between *C* and the focal point *F*. In addition, the **lateral magnification,** *m* (the ratio of the image's height to the object's), is less than one.

The characteristics of the images formed by a concave mirror are summarized in Table 8-1. Note that, when the object is beyond *F*, the image is inverted and real, and its size grows as the object gets closer to *F*.

[†] In general, incident rays parallel to the principal axis do not converge to a single point upon reflection. This is because of an effect called **spherical aberration**. However, if the curvature of the mirror is small, the rays will almost converge to a single point.

Copyright © by June Chung – All rights reserved. Reproduction of this publication without the written permission of the copyright owner is strictly prohibited.

Table 8-1: Image characteristics for a concave mirror

Position of object	Image characteristics			
	position	real/ virtual	upright/ inverted	m
beyond C	between C and F	real	inverted	< 1
at C	at C	real	inverted	$= 1$
between C and F	beyond C	real	inverted	> 1
at F	No image is formed			
in front of F	behind mirror	virtual	upright	> 1

On the other hand, when the object is in front of *F*, the image is upright and virtual, and its size gets smaller as the object approaches the mirror. The **ray diagram** in Fig. 8-5 shows the formation of this virtual image. Again, our brain imagines that the rays originated from behind the mirror, as was the case for a plane mirror. Because the virtual images formed by a concave mirror are upright and larger than the object, concave mirrors can be used to magnify objects and are used as makeup and shaving mirrors.

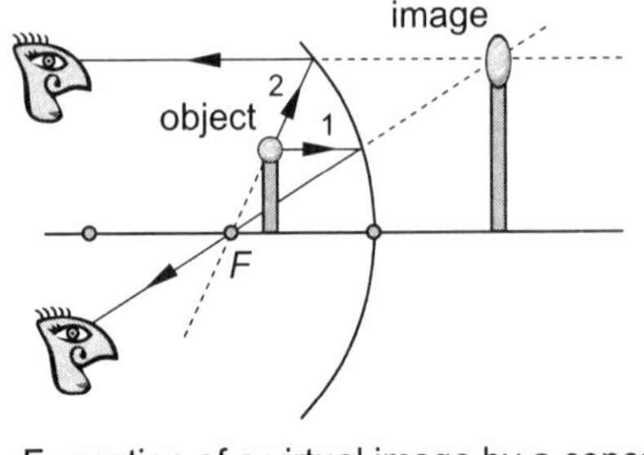

Figure 8-5: Formation of a virtual image by a concave mirror

Copyright © by June Chung – All rights reserved. Reproduction of this publication without the written permission of the copyright owner is strictly prohibited.

- For a concave mirror, the image of an object can be found by drawing a ray diagram according to these rules:
    1. Incident rays parallel to the principal axis pass through the focal point upon reflection.
    2. Incident rays passing through the focal point travel parallel to the principal axis upon reflection.
- The characteristics of the image formed by a concave mirror depend on how far the object is from the mirror. ⟹ See Table 8-1.

Example 8-3

Draw the ray diagram for an object situated between *C* and *F*. Confirm that an inverted, real image appears beyond *C* and its size is larger than the object.

Copyright © by June Chung – All rights reserved. Reproduction of this publication without the written permission of the copyright owner is strictly prohibited.

Key Point 8.5: Image Formation by a Convex Mirror

rules for drawing a ray diagram for convex mirrors

For a **convex mirror**, a ray diagram is constructed using these two rules:

1. Draw an incident ray parallel to the principal axis from the object to the mirror. Then the reflected ray travels as if it came straight from the focal point behind the mirror.
2. Draw an incident ray from the object towards the focal point behind the mirror. The reflected ray travels parallel to the principal axis.

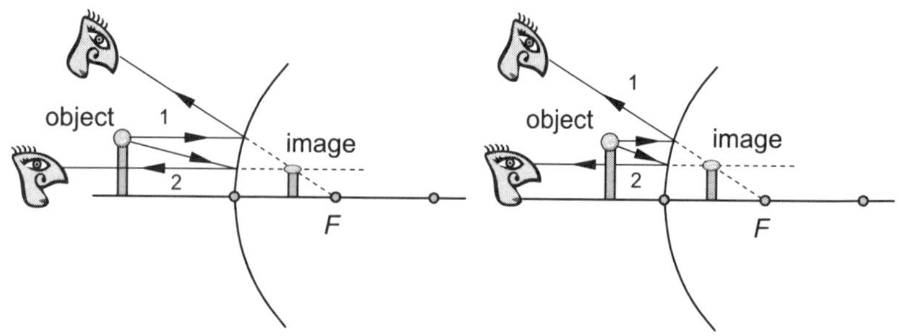

Figure 8-6: Image formation by a convex mirror

characteristics of a convex mirror image

Using the above rules, we can find the image of an object behind the mirror where these two reflected rays appear to meet, as shown in Fig. 8-6. As the rays do not converge in front of the mirror, it seems that they originate from behind the mirror, and we see a virtual image. This is always true for convex mirrors, and the image is always upright and smaller than the object. In fact, for convex mirrors, the location of the object does not affect the characteristics of the image. That is, wherever the object is located, the image always appears behind the mirror, and it is always upright and virtual. While its size is always smaller than the object (thus $m < 1$), it increases as the object approaches the mirror.

> - For a convex mirror,
>   1. Incident rays parallel to the principal axis are reflected as if they came straight from the focal point behind the mirror.
>   2. Incident rays traveling toward the focal point travel parallel to the principal axis upon reflection.
> - The images formed by convex mirrors are always upright, virtual, and smaller than the object.

Copyright © by June Chung – All rights reserved. Reproduction of this publication without the written permission of the copyright owner is strictly prohibited.

Key Point 8.6: The Mirror Equation and the Lateral Magnification

mirror equation

The **mirror equation** relates the object distance, d_o, and the image distance, d_i, to the focal length, f:

$$\frac{1}{d_o}+\frac{1}{d_i}=\frac{1}{f} \qquad \ldots (8.1)$$

In the equation, the sign of the image distance is <u>negative</u> if the image appears <u>behind</u> the mirror and is virtual. Similarly, the focal length of a convex mirror is <u>negative</u> because its focal point is <u>behind</u> the mirror. Since the object is always in front of the mirror, the object distance, d_o, is always positive.

example

For example, if a 30-cm tall object is 25 cm in front of a concave mirror whose focal length is 40 cm, the location of its image can be calculated as

$$\frac{1}{d_i}=\frac{1}{f}-\frac{1}{d_o}=\frac{1}{0.40}-\frac{1}{0.25}=-1.5\ \text{m}^{-1}\ ;\ d_i=-0.67\ \text{m}$$

Since the sign of d_i is negative, the image appears behind the mirror, which means it is a virtual image, and it is located 67 cm from the mirror.

lateral magnification

The **lateral magnification**, m, is defined as the ratio of the image's height, h_i, to the object's, h_o, and it is related to d_o and d_i as

$$m=\frac{h_i}{h_o}=-\frac{d_i}{d_o} \qquad \ldots (8.2)$$

The height of the image is considered <u>positive</u> if the image is <u>upright</u> and <u>negative</u> if it is <u>inverted</u>. (h_o is always positive.) Thus, the height of the image in the above example is

$$h_i=-h_o\frac{d_i}{d_o}=-(0.30)\frac{-0.67}{0.25}=0.80\ \text{m}$$

Because the sign of h_i is positive, the image is upright. The lateral magnification is 2.7, which means the image is 2.7 times larger than the object.

In summary, using only Eqs. (8.1) and (8.2), we can determine all the characteristics of the image formed by spherical mirrors without drawing a ray diagram. With the mirror equation in Eq. (8.1), we can find the location of an object's image and whether the image is real or virtual. With the magnification equation in Eq. (8.2), we can determine the size of the image and whether it is upright or inverted.

Copyright © by June Chung – All rights reserved. Reproduction of this publication without the written permission of the copyright owner is strictly prohibited.

- Mirror equation:

$$\frac{1}{d_o}+\frac{1}{d_i}=\frac{1}{f}$$

- Lateral magnification:

$$m=\frac{h_i}{h_o}=-\frac{d_i}{d_o}$$

- Put a negative sign for any distance measured behind the mirror.
- Put a negative sign for the height of the image if the image is inverted.
- Virtual images are always upright.

Example 8-4

An object is in front of a convex mirror whose focal length is 33.0 cm. If the image is 18.0 cm tall and 10.0 cm away from the mirror, find (a) whether the image is upright or inverted; (b) the image distance, d_i; (c) the object distance, d_o; and (d) the height of the object.

Example 8-5

Nancy, who is 160 cm tall, stands in front of a concave mirror and sees her own image, which is 40 cm tall and 12 cm in front of the mirror. Find (a) the distance between Nancy and the mirror; (b) the focal length of the mirror; (c) whether the image is real or virtual; and (d) whether the image is upright or inverted.

Copyright © by June Chung – All rights reserved. Reproduction of this publication without the written permission of the copyright owner is strictly prohibited.

Key Point 8.7: **Lenses**

lens

Lenses are made from transparent materials so that light rays can pass through them. They use the property of refraction to form an image, and they are either converging or diverging lens, depending on whether the refracted light rays converge or diverge. A magnifying glass and eyeglasses are an example of each type.

bi-convex and bi-concave lenses

The lenses we consider in this course are **bi-convex** and **bi-concave lenses**, where both faces form a section of a sphere. As shown in Fig. 8-7, bi-convex lenses are thicker in the middle, and bi-concave lenses are thinner in the middle.

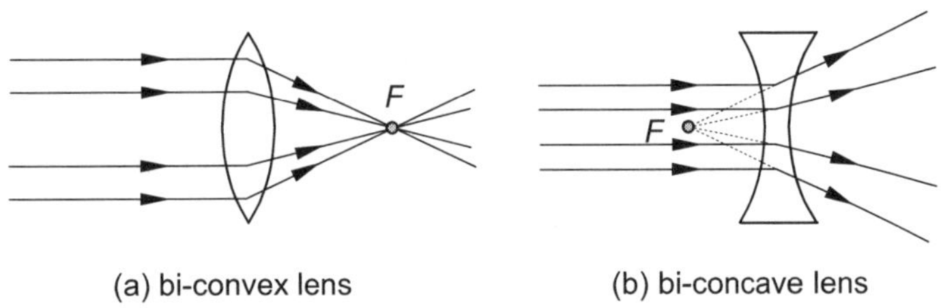

Figure 8-7: Refraction of light rays through thin lenses

anatomy of thin lenses

For these lenses, the principal axis is a straight line that divides the lens in half and is perpendicular to both faces, as shown in Fig. 8-8. The **vertical axis** is another line that divides the lens in half, but it is perpendicular to the principal axis. Unlike spherical mirrors, each lens has two focal points, one on each side. For a bi-convex lens, the main focal point is where incident rays parallel to the principal axis converge after refraction. For a bi-concave lens, it is the point from which the refracted rays

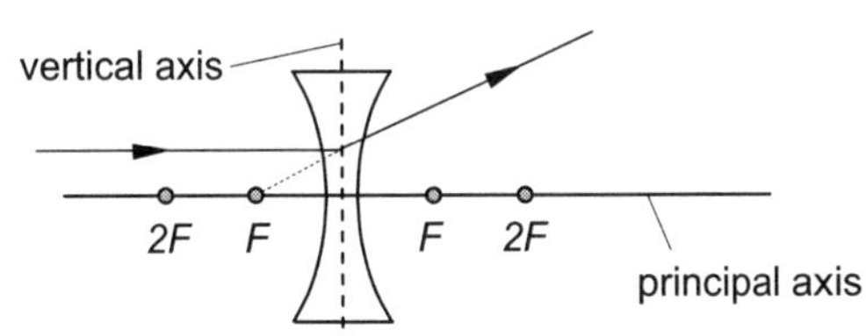

Figure 8-8: Anatomy of a bi-concave lens

Note: We know that light will refract twice through the lens, once when entering and again when leaving. However, if the lens is thin enough, we can assume refraction occurs once at the vertical axis. Note that maximum refraction occurs at the top and bottom of the lens because the angle of incidence is the greatest there. Rays traveling along the principal axis suffer no refraction as the angle of incidence is zero and they are perpendicular to the lens.

Copyright © by June Chung - All rights reserved. Reproduction of this publication without the written permission of the copyright owner is strictly prohibited.

appear to diverge when incident rays parallel to the principal axis meet the lens. The distance from the center of a lens to the focal point is called the focal length. Note that lenses do not have a center of curvature. Instead, there is an imaginary point on each side which we refer to as the 2*F* point, which is just like the center of curvature for spherical mirrors.

> - Light rays refracted through a bi-convex lens converge; those through a bi-concave lens diverge.
> - Each lens has two focal points, one on each side.
> - For a bi-convex lens, refraction causes incident rays parallel to the principal axis to converge to the focal point past the lens.
> - For a bi-concave lens, refraction causes incident rays parallel to the principal axis to diverge from the focal point before the lens.
> - Each lens has a pair of imaginary points referred to as the 2*F* points, one on each side of the lens.

Example 8-6

For a bi-convex lens, draw the principal and vertical axes and indicate the focal points. Then, for an incident ray parallel to the principal axis, draw the refracted ray.

Copyright © by June Chung – All rights reserved. Reproduction of this publication without the written permission of the copyright owner is strictly prohibited.

Key Point 8.8: **Image Formation by a Bi-Convex (Converging) Lens**

rules for drawing a ray diagram for bi-convex lens

There are two rules for drawing the ray diagram for a bi-convex lens, and they are similar to those for a concave mirror:

1. Draw an incident ray parallel to the principal axis from the object to the lens. Upon refraction, this ray passes through the focal point.
2. Draw an incident ray connecting the object, the focal point on the object's side, and the lens. Upon refraction, this ray travels parallel to the principal axis.

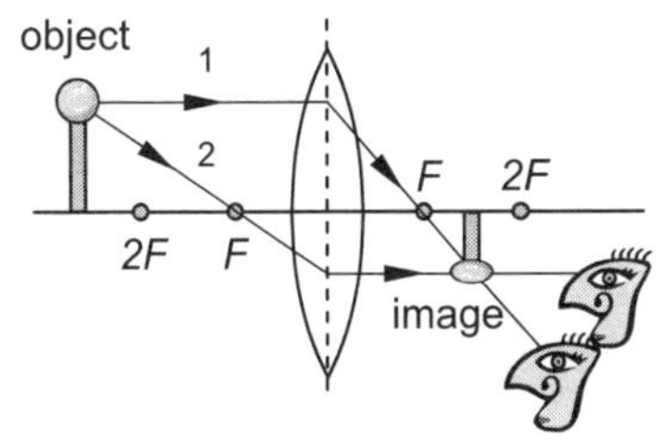

Figure 8-9: Image formation by a bi-convex lens

characteristics of a bi-convex lens image

An image is formed where the refracted rays meet, as shown in Fig. 8-9. If you try drawing ray diagrams for objects in various positions, you will see that the characteristics of the image depend on the object distance, d_o. In fact, the way the image changes with respect to the object distance is identical to that for a concave mirror (see Table 8-1, p. 162). Therefore, for an object located beyond the 2*F* point (which corresponds to the center of curvature for a mirror), an inverted, real image appears between 2*F* and the focal point, *F*. Keep in mind that a real image formed by a lens is on the opposite side of the lens from the object while that for a mirror appears in front of it.

magnifying glass as example

With a magnifying glass, you can confirm the characteristics of the images in Table 8-1. If you hold the magnifying glass close to, say, a newspaper, you can see upright, enlarged images of the letters behind the glass. This happens because the newspaper is within the focal length of the lens. If you start moving the glass away from the newspaper, the upright image gets bigger and bigger until it suddenly disappears when the newspaper is exactly at the focal point of the lens. Beyond that point, an inverted image will appear and decrease in size as you keep moving the lens away from the newspaper.

Copyright © by June Chung – All rights reserved. Reproduction of this publication without the written permission of the copyright owner is strictly prohibited.

> - For a bi-convex lens, the image of an object can be found by drawing a ray diagram according to these rules:
>   1. Incident rays parallel to the principal axis pass through the focal point upon refraction.
>   2. Incident rays passing through the focal point travel parallel to the principal axis upon refraction.
> - The relation between image characteristics and the object distance for a bi-convex lens is identical to that for a <u>concave</u> mirror.
>
>   ⟹ See Table 8-1 (p. 164).

Example 8-7

Draw the ray diagram for an object situated between a bi-convex lens and its focal point, *F*. Confirm that the image is upright, virtual, and larger than the object.

Copyright © by June Chung – All rights reserved. Reproduction of this publication without the written permission of the copyright owner is strictly prohibited.

Key Point 8.9: Image Formation by a Bi-Concave (Diverging) Lens

rules for drawing a ray diagram for bi-concave lens

A ray diagram for a bi-concave lens is constructed using these two rules:

1. Draw an incident ray parallel to the principal axis from the object to the lens. Then this ray, upon refraction, travels as if it came from the focal point in front of the lens (that is, on the object's side of the lens).
2. Draw an incident ray from the object towards the focal point on the opposite side of the lens. Then this ray, upon refraction, travels parallel to the principal axis.

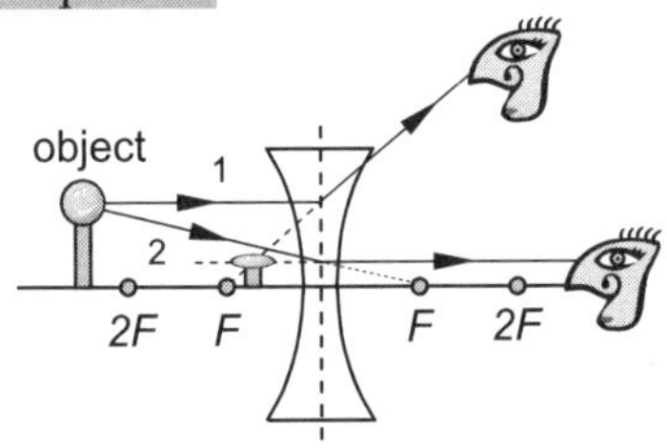

Figure 8-10: Image formation by a bi-concave lens

characteristics of a bi-concave lens image

The image forms where these two refracted rays appear to meet, as shown in Fig. 8-10. Just as the reflected rays do not converge for a convex mirror, the rays refracted through a bi-concave lens do not converge. Thus, we see a virtual image just as we did with a convex mirror. In fact, the bi-concave lens and the convex mirror share the same image characteristics; the image is always upright, virtual, and smaller than the object. Furthermore, the location of the object does not affect the characteristics of the image.

example

The corrective lenses prescribed for nearsightedness (myopia) are diverging lenses. Thus, we always see upright images through those lenses although they are smaller than the actual objects.

> ➢ For a bi-concave lens,
> 1. Incident rays parallel to the principal axis are refracted as if they originated from the focal point on the object's side.
> 2. Incident rays traveling toward the focal point on the opposite side travel parallel to the principal axis upon refraction.
>
> ➢ Bi-concave lenses and convex mirrors share the same image characteristics ⇒ see Key Point 8.5.

Copyright © by June Chung – All rights reserved. Reproduction of this publication without the written permission of the copyright owner is strictly prohibited.

Key Point 8.10: The Lens Equation

lens equation

The **lens equation** is essentially the same as the mirror equation:

$$\frac{1}{d_o}+\frac{1}{d_i}=\frac{1}{f} \quad \ldots (8.3)$$

In the equation, the sign of the image distance is negative if the image appears on the same side of the lens as the object (that is, the image is virtual). As for the focal length, it is positive for a bi-convex lens and negative for a bi-concave lens. The object distance is always positive.

lateral magnification

The lateral magnification, m, for a lens is calculated in the same way as that for a mirror:

$$m=\frac{h_i}{h_o}=-\frac{d_i}{d_o} \quad \ldots (8.4)$$

The height of the image is considered positive if the image is upright and negative if it is inverted. (h_o is always positive.)

solving lens problems

When solving problems involving lens, you can always find the answers by applying Eqs. (8.3) and (8.4). The trick to solving this kind of problem is choosing the correct signs for the quantities in these equations, so pay attention to the signs. After you get an answer from the equations, you can confirm it by drawing a ray diagram using the rules for that lens.

- Lens equation:

$$\frac{1}{d_o}+\frac{1}{d_i}=\frac{1}{f}$$

- Lateral magnification:

$$m=\frac{h_i}{h_o}=-\frac{d_i}{d_o}$$

- Put a negative sign for image distance if the image appears on the object's side of the lens.
- The focal length of a bi-convex lens is positive; that of a bi-concave lens is negative.
- Put a negative sign for the height of the image if the image is inverted.
- Virtual images are always upright.

Copyright © by June Chung - All rights reserved. Reproduction of this publication without the written permission of the copyright owner is strictly prohibited.

Example 8-8

The human eye consists of a cornea, iris, lens, and retina, as shown below. The cornea-lens system is shaped like a converging lens and thus can refract light rays to create images on the retinal surface. In general, the focal length of the cornea-lens system is about 1.80 cm. If a 1.75-m tall person stands 1.05 m from our eyes, determine (a) the image distance and whether the image is real or virtual; and (b) the image size and whether the image is upright or inverted.

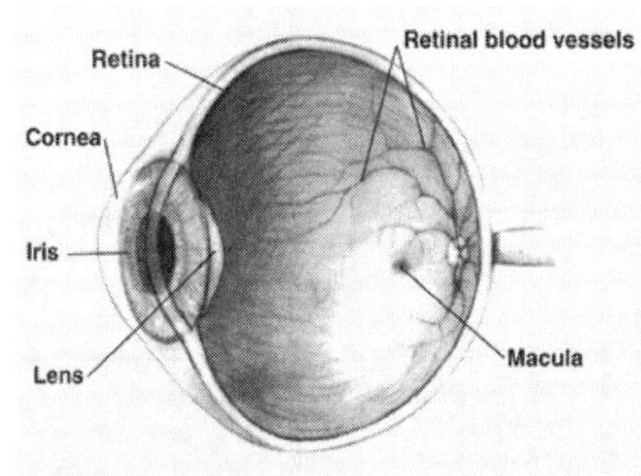

[Source: http://en.wikipedia.org/wiki/Image:Human_eye_cross-sectional_view_grayscale]

Copyright © by June Chung - All rights reserved. Reproduction of this publication without the written permission of the copyright owner is strictly prohibited.

ANSWER KEY TO EXAMPLES

Example 8-1

The diagrams drawn below show the location of each image. If possible, stand in front of two mirrors joined at a right angle and confirm that you do see three images of yourself.

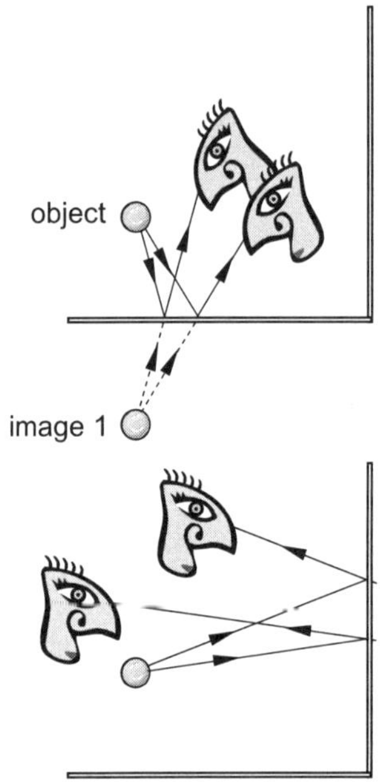

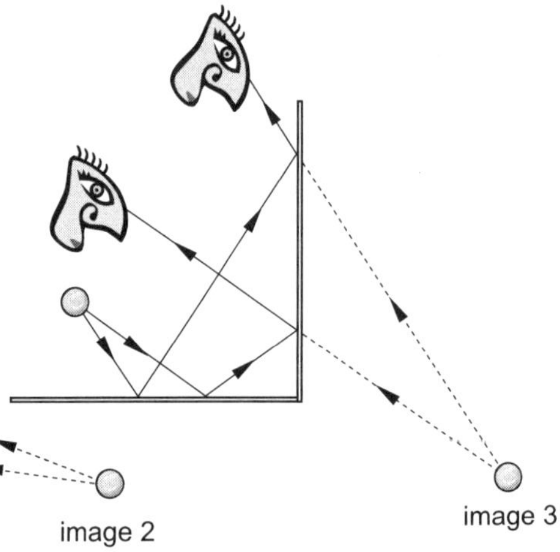

Example 8-2

The principal axis and the three points (*V*, *P*, *C*) are shown below. Since the focal length is one-half the radius of the sphere, it is 0.22 m.

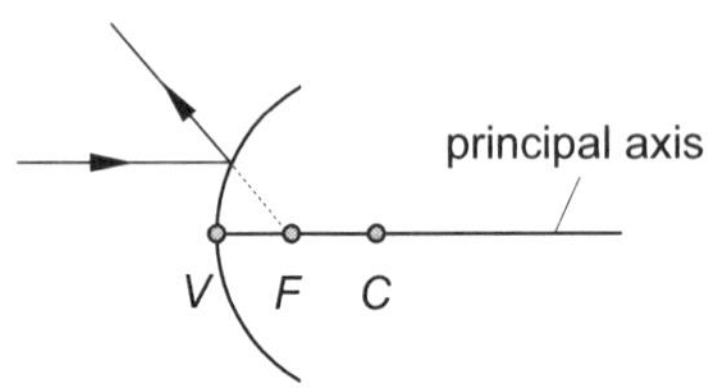

Example 8-3

The ray diagram is shown below:

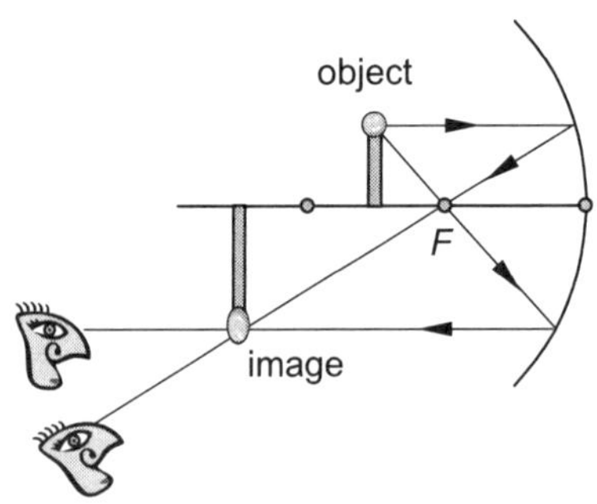

Copyright © by June Chung – All rights reserved. Reproduction of this publication without the written permission of the copyright owner is strictly prohibited.

Example 8-4

(a) Images formed by a convex mirror are always upright. Thus, $h_i = +0.180$ m.

(b) Images formed by a convex mirror are always virtual and appear behind the mirror. Thus, $d_i = -0.100$ m:

(c) Use the mirror equation to find the object distance. Since the focus is behind the mirror, $f = -0.330$ m.

$$\frac{1}{d_o} = \frac{1}{f} - \frac{1}{d_i} = \frac{1}{-0.330} - \frac{1}{-0.100} = 6.97 \text{ m}^{-1};\ d_o = 0.143 \text{ m}$$

Thus, the object is 14.3 cm in front of the mirror.

(b) Use Eq. (8.2) to find h_o:

$$\frac{0.180}{h_o} = -\frac{-0.100}{0.143}$$

$$h_o = 25.8 \text{ cm}$$

Hence, the object is 25.8 cm tall.

Example 8-5

The known quantities are the height of the object ($h_o = 1.60$ m), the height of the image ($h_i = 0.40$ m), and the image distance ($d_i = 0.12$ m). Because d_o and h_o are always positive, it means that h_i is in fact negative.

(a) Substitute the known values into Eq. (8.2):

$$\frac{-0.40}{1.60} = -\frac{0.12}{d_o}$$

Solving for d_o, we find that Nancy is standing 48 cm in front of the mirror.

(b) Use the mirror equation to find the focal length:

$$\frac{1}{f} = \frac{1}{d_o} + \frac{1}{d_i} = \frac{1}{0.48} + \frac{1}{0.12} = 10.4 \text{ m}^{-1};\ f = 0.096 \text{ m}$$

Thus, the focal length is 9.6 cm.

(c) Because the image appears in front of the mirror, the image is real.

(d) The image is inverted because h_i is negative.

Example 8-6

The two axes and the focal points are drawn below. The incident ray, when refracted through the lens, passes through the focal point at right as shown.

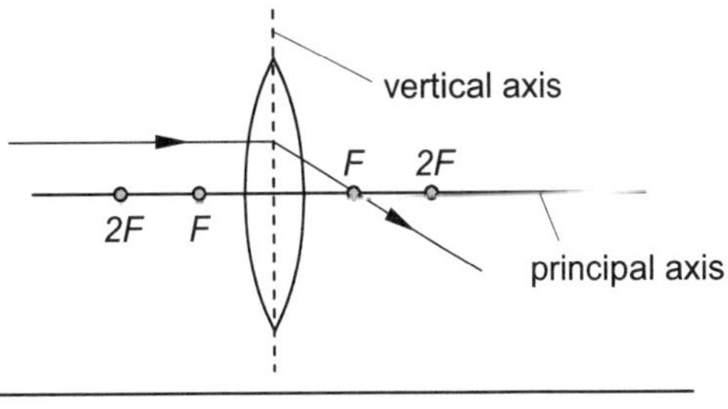

Copyright © by June Chung – All rights reserved. Reproduction of this publication without the written permission of the copyright owner is strictly prohibited.

Example 8-7

Draw a ray diagram, as shown below, using the two rules for a bi-convex lens. The refracted rays never converge, and we think that the refracted rays came from behind the lens. As a result, a virtual image is formed.

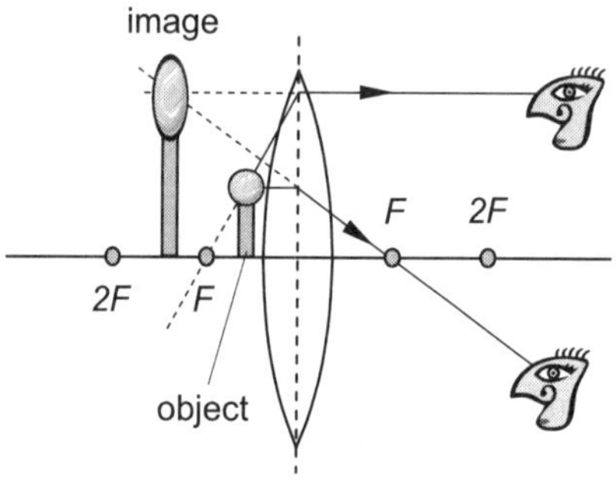

Example 8-8

(a) The focal length of a converging lens is positive. Use the mirror equation to find the image distance:

$$\frac{1}{d_i}=\frac{1}{f}-\frac{1}{d_o}=\frac{1}{0.018}-\frac{1}{1.05}=54.6\ \text{m}^{-1};\quad d_i=0.0183\ \text{m}$$

Thus, the image distance is 1.83 cm. Since d_i is positive, it means that the image is real.

(b) Use the definition of the lateral magnification in Eq. (8.4) to find the image size:

$$\frac{h_i}{1.75}=-\frac{0.0183}{1.05}$$

$$h_i=-3.05\ \text{cm}$$

Therefore, the image is 3.05 cm tall, and it is inverted because it is negative.

Copyright © by June Chung – All rights reserved. Reproduction of this publication without the written permission of the copyright owner is strictly prohibited.

PROBLEMS ON CHAPTER 8: GEOMETRIC OPTICS

(Answer Key on p. 234)

1 – 10. Fill in the blanks

1. The _________ model of light assumes that light rays travel in straight lines.
2. A _________ image has no light rays actually coming through it.
3. The _________ point of a mirror is located halfway between the center of curvature and the vertex.
4. Mirrors form images by _________ incident light rays.
5. The ratio of the image's height to the object's height is called the _______ _______.
6. For mirrors, the sign of the image distance, d_i, is _________ if the image appears in front of the mirror.
7. The sign of the image height, h_i, is negative if the image is _________.
8. _________ lenses converge incident light rays.
9. For lenses, _________ images appear on the opposite side from the object.
10. The _________ equation relates the focal length of a lens with the image distance and the object distance.

11 – 20. True/False questions

11. _____ An image formed by a plane mirror is smaller than the size of the object.
12. _____ Images formed by a plane mirror are always virtual images.
13. _____ A light ray passing through the focal point of a concave mirror, when reflected, travels parallel to the principal axis.
14. _____ When an object is located between the focal point and the vertex of a concave mirror, no image is formed.
15. _____ Images formed by a convex mirror are always upright images.
16. _____ As an object approaches a convex mirror, its image increases in size.
17. _____ A magnifying glass is a converging lens.
18. _____ The characteristics of images formed by a bi-convex lens are identical to those of images formed by a convex mirror.
19. _____ Images formed by a diverging lens are always bigger in size than the object.
20. _____ Thin lenses do not have the center of curvature as spherical mirrors do.

Copyright © by June Chung – All rights reserved. Reproduction of this publication without the written permission of the copyright owner is strictly prohibited.

21 – 25. Multiple choice questions

21. Which of the following is true for images formed by a concave mirror?

    a) They are always virtual images.
    b) They are always upright images.
    c) They are always bigger in size than the objects.
    d) They disappear when the objects are at the focal point of the mirror.

22. Which of the following is false for images formed by a convex mirror?

    a) They are always virtual images.
    b) They are always upright images.
    c) They are always bigger in size than the objects.
    d) They are always within the focal length behind the mirror.

23. An object is located two focal lengths away from a bi-convex lens. Which of the following is true for the image of the object?

    a) The image is half the size of the object.
    b) The image is real.
    c) The image is upright.
    d) No image is formed.

24. Which of the following is false for a bi-concave lens?

    a) The lens diverges incident light rays.
    b) The image characteristics are identical to those for a convex mirror.
    c) The images are always virtual, inverted, and smaller in size than the object.
    d) The focal length is negative.

25. If the image formed by a concave mirror is on the same side of the mirror as the object, which of the following statements is true?

    a) The image is inverted.
    b) The image is virtual.
    c) The image is farther from the mirror than the object.
    d) The object is behind the center of curvature.

Copyright © by June Chung – All rights reserved. Reproduction of this publication without the written permission of the copyright owner is strictly prohibited.

26 – 36. Written-response questions

26. Randy, who is 172 cm tall, wants to see the full image of himself (from the top of his head to his feet) using a plane mirror. Find the minimum required length of the mirror. (His eyes are 12 cm below the top of his head.)

27. By drawing a ray diagram, find the image of the object located 7.3 cm in front of a convex mirror whose focal length is 5.0 cm.

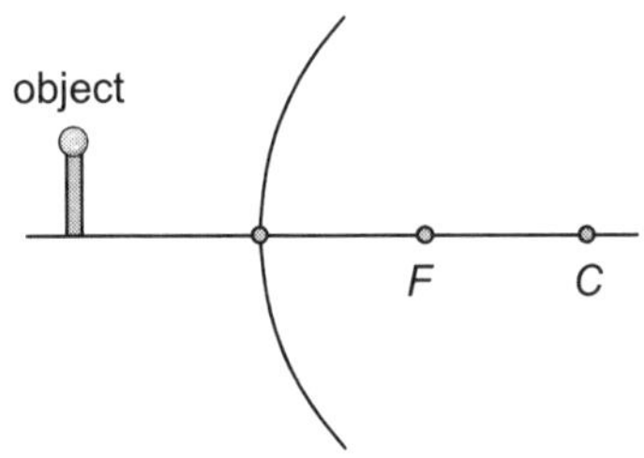

28. If the height of the image in Problem 27 is 0.87 m, find the image distance and the height of the object.

29. A 0.2-cm tall object is placed at the focal point of a concave mirror whose focal length is 8 cm. Find the size of the image.

30. A 40-cm object is placed 25 cm in front of a spherical mirror. If a real image appears 5.0 cm from the mirror, find (a) whether the mirror is concave or convex, (b) the focal length of the mirror, and (c) the height of the image.

31. A large convex mirror is installed in front of "Miniature World" so that people on the street can look at the small images of themselves. One day, a robber stole many expensive items from the shop and ran away. A few days later, a suspect was caught, but he insisted he was innocent and claimed that he had actually witnessed the event that day, describing what he saw as follows: "As he ran away from the store, his reflected image on the mirror got smaller. But I could not apprehend that guy because the image soon became inverted." Based on his testimony, is he believable? Explain.

Copyright © by June Chung – All rights reserved. Reproduction of this publication without the written permission of the copyright owner is strictly prohibited.

32. By drawing a ray diagram, find the image of the object located 1.2 cm in front of a bi-convex lens whose focal length is 2.7 cm.

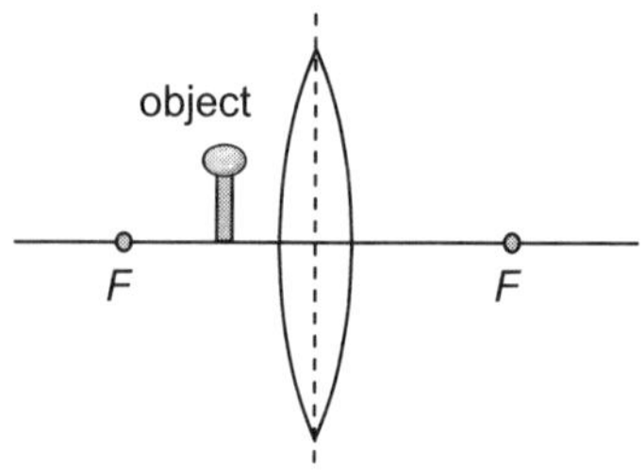

33. If the height of the object in Problem 32 is 1.8 m, find the image distance and the height of the image.

34. When a 30-cm tall object is 2.2 m away from a bi-concave lens, the image is formed 1.2 m away from the lens. When the object is moved to a new location, its image now appears 1.7 m away from the lens. Find the object distance and the height of the image.

35. When a 3.5-cm tall object is 22 cm away from a lens, a 2.0-cm tall virtual image is formed. Find the focal length of the lens, and determine the type of lens used to form the virtual image.

36. A bi-convex lens whose focal length is 1.40 cm is placed 2.50 cm in front of a concave mirror whose focal length is 0.150 cm. If a 2.20-cm tall object is placed 3.60 cm in front of the bi-convex lens as shown, find (a) the characteristics of the image formed by the lens and (b) the size of the final image formed by the concave mirror and the image distance. If we use a bi-concave lens with the same focal length instead of the bi-convex lens, would we be able to obtain a reflected image from the concave mirror? Explain.

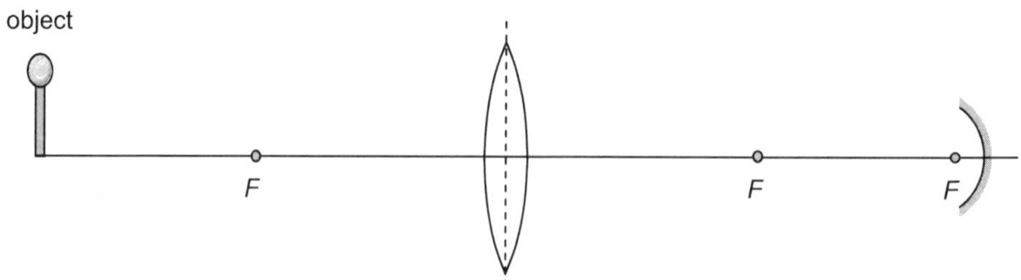

Copyright © by June Chung – All rights reserved. Reproduction of this publication without the written permission of the copyright owner is strictly prohibited.

CHAPTER 9

RELATIVITY

So far we have been studying Newtonian physics, which enables us to explain everyday phenomena. But Newtonian physics cannot explain the motion of very tiny particles like electrons, or the movement of objects moving at speeds approaching the speed of light. For this we need the work of Einstein and his theories of relativity. Einstein's ideas were revolutionary. He proposed that mass and energy are different forms of the same thing, so that mass can be converted into energy and vice versa. He said that the speed of light is the only absolute. Length and time, long believed to have absolute values, depend on the relative motion of the object and the observer. These ideas are difficult to grasp, but they provide the basis for understanding nuclear energy and the global positioning systems that we use today.

In this chapter, you will study:

- ✓ the principle of relativity
- ✓ Maxwell's theory of electromagnetism and the Michelson-Morley experiment
- ✓ postulates of the special theory of relativity
- ✓ simultaneity
- ✓ time dilation, length contraction, and mass increase
- ✓ $E = m\,c^2$

Copyright © by June Chung - All rights reserved. Reproduction of this publication without the written permission of the copyright owner is strictly prohibited.

Key Point 9.1: The Principle of Relativity

laws of physics in inertial frame of reference

When an observer is either at rest or moving at constant velocity, we say that events occur in an **inertial frame of reference**.[†] Although the apparent motion of an object depends on the relative motion of the observer, the same set of laws can be applied by different observers, as long as the observers themselves are not accelerating.

example

For example, suppose you drop a ball inside a train moving at constant velocity. For a passenger on the train, the ball would appear to fall vertically. However, for an observer standing on the ground outside the train, the ball would appear to travel in a curved path, as illustrated in Fig. 9-1. Kinematic equations (p. 33) can be used in both cases because neither observer is accelerating, and each description of the ball's motion is equally valid.

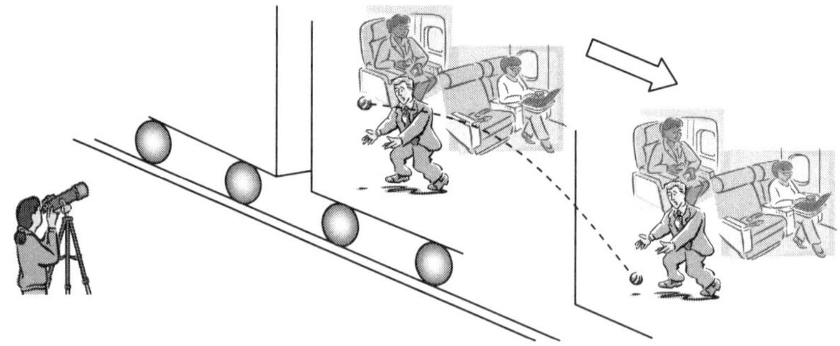

Figure 9-1: Trajectory of a dropped ball as seen by an observer on the ground

principle of relativity

Hence, it is concluded that no one inertial frame of reference is more special than another. This is the **principle of relativity**, which states that the basic laws of physics are the same in all inertial frames of reference.

- Inertial frame of reference is a reference frame in which the observer is not accelerating.
- The laws of physics are the same in all inertial frames of reference.
- No one inertial frame of reference is more special than another.

[†] An inertial frame of reference is defined to be a reference frame in which Newton's laws of motion are valid. All physical laws covered thus far in this book are valid in inertial frames of reference.

Copyright © by June Chung - All rights reserved. Reproduction of this publication without the written permission of the copyright owner is strictly prohibited.

Key Point 9.2: Maxwell's Theory of Electromagnetism and the Michelson-Morley Experiment

Maxwell's discovery of speed of light

In 1865 a Scottish scientist named James Maxwell showed that electromagnetic waves travel at the speed of light, which he calculated to be 3.00×10^8 m/s in a vacuum. This discovery puzzled scientists at the time because speed must be relative to the motion of the observer. For instance, to an observer traveling at the speed of light, electromagnetic waves would appear to be at rest when both the observer and the waves are traveling in the same direction. This reasoning seemed to suggest that there exists a special frame of reference in which electromagnetic waves can travel at 3.00×10^8 m/s, and they tried to find that absolute frame.

Michelson-Morley experiment

Many attempts were made to solve this mystery, and the experiment done by Michelson and Morley in 1887 is perhaps the most significant. As most scientists then believed, Michelson and Morley thought that electromagnetic waves traveled through a medium called the "ether," which was thought to permeate all space. They reasoned that, because of the earth's rotation, there would be an "ether wind," and electromagnetic waves traveling in different directions with respect to this ether wind would travel at different speeds. With the experimental setup shown in Fig. 9-2, they tried to measure a difference in the speed of light in different directions but failed to detect any. Unable to explain the result of this experiment, scientists had to wait until 1905, when Einstein proposed a new theory of relativity that drastically altered the classical concepts of space, time, and mass.

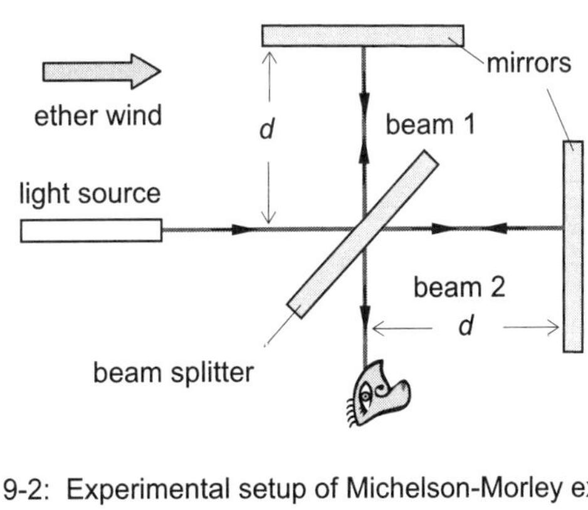

Figure 9-2: Experimental setup of Michelson-Morley experiment

Copyright © by June Chung – All rights reserved. Reproduction of this publication without the written permission of the copyright owner is strictly prohibited.

- Maxwell's theory shows that all electromagnetic waves travel at the speed of light, which is 3.00×10^8 m/s in a vacuum.
- Michelson and Morley tried to detect a difference in the speed of light traveling in different directions with respect to the ether but failed to detect any difference at all.

Copyright © by June Chung – All rights reserved. Reproduction of this publication without the written permission of the copyright owner is strictly prohibited.

Key Point 9.3: Postulates of the Special Theory of Relativity

two postulates of the special theory of relativity

One of the papers Einstein published in 1905 introduced his **special theory of relativity**, which is based on the following two postulates:

1. The laws of physics are the same in all inertial frames of reference.
2. The speed of light is the same in all inertial frames of reference.

The first postulate is the principle of relativity, covered in Key Point 9.1, and is easy to understand. Scientists before Einstein were also aware of this principle. On the other hand, in order to understand the second postulate, one has to give up the classical concept that space and time are 'absolute.'

the second postulate of the special theory of relativity

In classical physics, the lengths of objects stay the same and time elapses at the same rate regardless of which reference frame you use. However, Einstein asserted in his special theory that space and time are not absolute. For example, two observers in different frames of reference may measure the length of an identical object and come up with different results. In addition, they may measure the time of an identical event and obtain different results. According to Einstein's special theory, the speed of light is the only quantity whose value is absolute. In other words, no matter who measures the speed of light, it will always be 3.00×10^8 m/s in a vacuum. Even to someone traveling at 1.00×10^8 m/s relative to the ground, the speed of light will be 3.00×10^8 m/s, not 4.00×10^8 m/s or 2.00×10^8 m/s or anything in between, which is what our everyday experience would lead us to expect. Although this may seem odd and counterintuitive, the special theory of relativity has been validated by many experimental observations.

> - Postulates of the special theory of relativity:
>   1. The laws of physics are the same in all inertial reference frames.
>   2. The speed of light is the same in all inertial reference frames.
>
> According to Einstein's special theory...
> - Space and time are not absolute.
> - The speed of light is absolute.

Copyright © by June Chung - All rights reserved. Reproduction of this publication without the written permission of the copyright owner is strictly prohibited.

Key Point 9.4: Simultaneity

act of observing objects or events

The act of observing an object requires that light emitted or reflected from the object reach our eyes. Because the speed of light is finite, it takes time for light to travel from an object to us. This means that the images of objects we see around us are in fact those of objects that existed some time ago. For instance, the image of the Sun we see in the sky is the image of the Sun that existed about 8 minutes ago because that is the time light takes to travel from the Sun to the earth.

example

From this and the fact that the speed of light is absolute, it is possible to show that two events that appear to occur simultaneously to one observer may not appear simultaneous to another observer. Consider the following example. One observer is standing an equal distance away from two points, as shown in Fig. 9-3. If lightning strikes simultaneously at A and B, light rays from A and B will reach the observer at the same time, and the observer will conclude that the lightning struck A and B simultaneously.

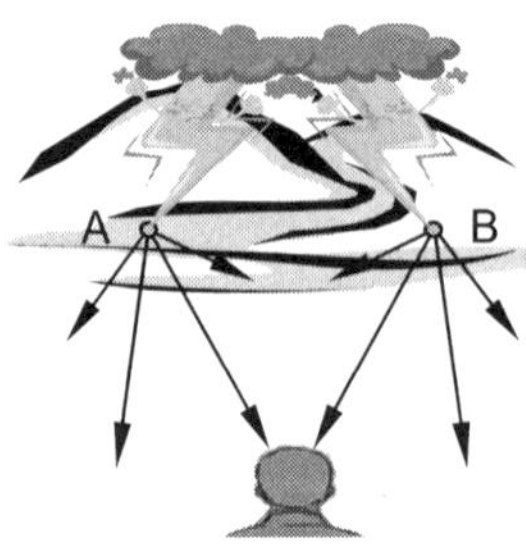

Figure 9-3: Lightning bolts seen by a stationary observer

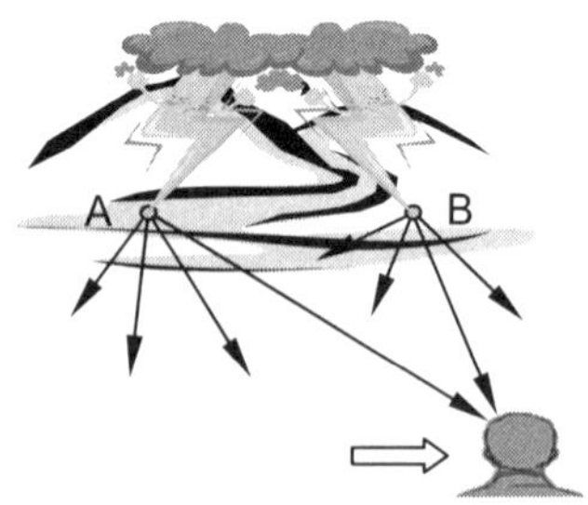

Figure 9-4: Lighting bolts seen by a moving observer

However, another observer moving at constant velocity, as in Fig. 9-4, will see the lightning at B before seeing the one at A, because the speed of light is not affected by the motion of the observer and light from B travels a shorter distance than light from A. Thus, the moving observer will claim that the lightning did not strike A and B simultaneously.

simultaneity

As shown in the example above, whether two events occur simultaneously or not depends on the motion of the observer, or the frame of reference. According to the special theory of relativity, no one frame of

Copyright © by June Chung – All rights reserved. Reproduction of this publication without the written permission of the copyright owner is strictly prohibited.

reference is more special than another. Hence, **simultaneity** is a relative concept, and it is impossible to judge which observer is correct.

> - For us to be able to see objects, light must travel at a finite speed from them to our eyes.
>   ⇨ We see the images of objects that existed some time ago.
> - The speed of light is constant and is independent of the observer's motion.
> - Whether two events occur simultaneously or not depends on the observer's motion.

Copyright © by June Chung – All rights reserved. Reproduction of this publication without the written permission of the copyright owner is strictly prohibited.

Key Point 9.5: Time Dilation

thought experiment on time dilation

Let's consider the following experiment. Inside a spaceship moving past Earth at high speed, a beam of light is emitted by a light source fixed to the top of the spaceship and travels across the ship to a receiver fixed to the bottom. To an observer aboard the spaceship, the path of the beam will be a vertical line. To an observer on Earth, however, the path will be a slanted line, as shown in Fig. 9-5 (b). (Notice the similarity between this and the example in Key Point 9.1)

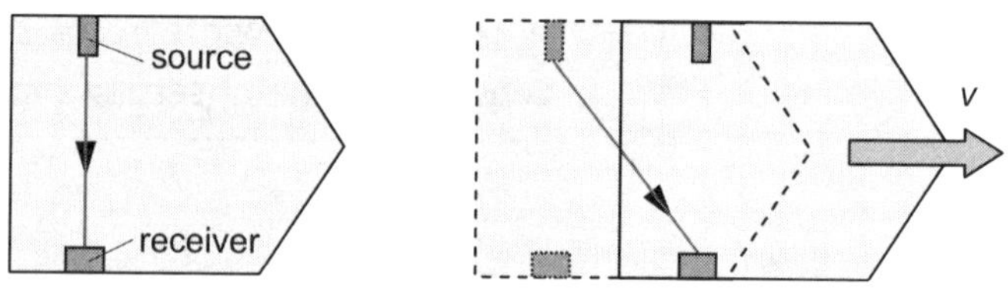

(a) to observer aboard spaceship (b) to observer on Earth

Figure 9-5: Path of light inside a spaceship moving at high speed

time dilation

While the speed of light is the same to both observers, the beam travels longer distance when viewed by an observer on Earth. Therefore, to the observer on Earth, it takes longer time for the beam to travel from the source to the receiver! This demonstrates that, for an identical event, time elapses at a different rate depending on the motion of the observer. It is easy to imagine that, the greater the speed of the spaceship, the farther the beam has to travel to get to the receiver. So, an event that occurs within 0.1 second to an observer aboard the ship may last 10 seconds to an observer on Earth. If there was a clock aboard the ship, it would appear to run more slowly than one on Earth (that is, to an observer on Earth). This phenomenon is known as **time dilation**. In general, the time taken by an event occurring inside a moving reference frame can be obtained from

$$\Delta t = \frac{\Delta t_o}{\sqrt{1 - \frac{v^2}{c^2}}} \qquad \ldots (9.1)$$

where Δt is the time interval measured by an observer at rest, Δt_o is the time interval measured by an observer inside the moving reference

Copyright © by June Chung - All rights reserved. Reproduction of this publication without the written permission of the copyright owner is strictly prohibited.

frame, v is the speed of the reference frame, and c is the speed of light, which is 3.00×10^8 m/s.

time dilation is relative

Note in the above example that, to an observer aboard the spaceship, it is <u>Earth</u> that is moving at high speed. Therefore, to that observer, events occurring in the spaceship appear normal while those occurring on Earth appear to unfold more slowly. To an observer on Earth, on the other hand, the opposite is true: events occurring in the spaceship appear to unfold more slowly while those occurring on Earth appear normal.

> - The rate at which time elapses depends on the observer's motion.
> - Time dilation: Moving clocks appear to run more slowly than a clock at rest.
> - As the speed of a moving reference frame increases, the clock in that reference frame runs more slowly.

Example 9-1

When a spaceship traveling at a speed of 2.00×10^8 m/s flies past Earth, the clock inside the spaceship shows 11 h 32 m 00.00 s. An observer on Earth waits for 2.00 seconds on his watch, and then looks at the clock inside the spaceship. What time is the clock showing at that instant?

Copyright © by June Chung – All rights reserved. Reproduction of this publication without the written permission of the copyright owner is strictly prohibited.

Key Point 9.6: Length Contraction

thought experiment on length contraction

Suppose a spacecraft travels from one planet to another at a constant speed, v. If the distance between the planets as measured by an observer on Earth is L_o, the time required for the trip can be obtained from

$$\Delta t = \frac{L_o}{v} \qquad \dots (9.2)$$

Δt is the time interval measured by the observer on Earth.

Now, observers on Earth think that clocks run more slowly inside the spacecraft than on Earth. Consequently, they think that the time spent on the trip is shorter when measured by an astronaut aboard the spacecraft. The time interval measured by the astronaut would be

$$\Delta t_o = \frac{L}{v} \qquad \dots (9.3)$$

Since Δt_o is less than Δt, L (the distance measured by the observer aboard the spacecraft) must be smaller than L_o (the distance measured by the observer on Earth). Using Eq. (9.1), we can derive the following relation between L and L_o:

$$L = L_o\sqrt{1-\frac{v^2}{c^2}} \qquad \dots (9.4)$$

length contraction

This equation shows that, to the astronaut, the distance between the planets appears shorter as well as everything between the planets (Fig. 9-6 (a)). On the other hand, to an observer on Earth, it is the spacecraft that appears shorter than normal instead of the distance between the planets (Fig. 9-6 (b)). This is called **length contraction**, and it occurs only along the direction of travel. In short, the length of an object measured while it is moving (relative to the observer) is shorter than that measured while it is at rest.

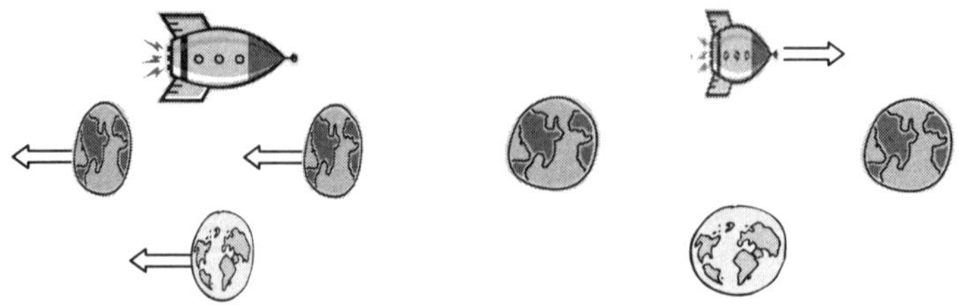

(a) to observer aboard spacecraft (b) to observer on Earth

Figure 9-6: Length contraction

Copyright © by June Chung – All rights reserved. Reproduction of this publication without the written permission of the copyright owner is strictly prohibited.

> - The length of an object depends on its motion relative to the observer.
> - Length contraction: The length of an object measured while it is moving relative to the observer is shorter than that measured when it is at rest.
> - The greater the speed of the object relative to the observer, the greater the length contraction.

Example 9-2

A spaceship travels past Earth at a speed of 6.00×10^7 m/s. An observer on Earth measures the length of the ship from its front to back and obtains 25.5 m. What is the length of the spaceship to an observer aboard the ship? If the observer on Earth is holding a meter stick parallel to the direction of the ship's motion, what is the length of the stick to the observer aboard the ship?

Copyright © by June Chung – All rights reserved. Reproduction of this publication without the written permission of the copyright owner is strictly prohibited.

Key Point 9.7: **Mass Increase**

mass increase

An event occurring in a reference frame that is moving relative to an observer appears to last longer, and objects in such a frame appear shorter in length along the direction of their motion. In addition, an object moving relative to an observer gains mass according to

$$m = \frac{m_o}{\sqrt{1 - \frac{v^2}{c^2}}} \qquad \ldots (9.5)$$

where m is the object's mass while it is moving and m_o is its mass measured when it is at rest (thus, m_o is called **rest mass**). The equation shows that, as an object is accelerated to a higher speed, the object gains mass.

mass increase is relative

For example, suppose an astronaut whose mass is 60 kg on Earth travels in a spaceship moving at high speed. To observers on Earth, the astronaut's mass seems to be more than 60 kg because, to them, she is the one who is moving. As far as the astronaut is concerned, however, she is at rest. It is the earth that is moving. Thus, to the astronaut, her mass in the spaceship will still be 60 kg, and everybody on Earth will appear to have gained mass.

- Mass increase: The mass of an object measured while it is moving relative to the observer is greater than that measured when it is at rest.
- The greater the speed of the object relative to the observer, the greater the mass increase.

Example 9-3

An astronaut is aboard a spaceship traveling at half the speed of light. An observer on Earth measures the mass of the moving spaceship to be 3.70×10^5 kg. What is the ship's mass when it returns to Earth?

Copyright © by June Chung – All rights reserved. Reproduction of this publication without the written permission of the copyright owner is strictly prohibited.

Key Point 9.8: $E = mc^2$

work-energy principle revisited

The net work done on an object and the change in its kinetic energy are related through the work-energy principle (see Key Point 5.3, p. 93):

$$W_{net} = \Delta E_K = \frac{1}{2}m\left(v_f^2 - v_i^2\right) \quad \ldots (9.6)$$

If the net work on an object is positive, the object gains kinetic energy. If it is negative, the object loses kinetic energy. In classical physics, mass is an absolute physical quantity. Thus, changes in kinetic energy could only be detected through changes in velocity.

work-energy principle in view of modern physics

As shown in the previous Key Point, however, the mass of an object depends on its motion relative to the observer. To a stationary observer, the mass of an object increases as its speed increases. This implies that, when an object gains kinetic energy, it gains not only speed but also mass! This leads to the idea that mass is a form of energy. The relationship between mass and energy is expressed as

$$E = mc^2 = m_o c^2 + E_K \quad \ldots (9.7)$$

where E is the total energy and $m_o c^2$ is the **rest energy** (that is, energy when the mass is at rest). Thus, when there is a change in an object's mass, its total energy also changes according to

$$\Delta E = (\Delta m)c^2 \quad \ldots (9.8)$$

equivalence of mass and energy

The equation shows that a considerable amount of energy can be released from even a small loss of mass because it is multiplied by c^2. For instance, one microgram of mass is converted into about 100 billion Joules of energy! Usually, such exchanges between mass and energy only occur during nuclear decays on a subatomic level (to be covered in Chapter 10).

> - A gain in kinetic energy of a body increases not only its speed but also its mass.
> - Mass is a form of energy and can be converted into its energy equivalent according to $\Delta E = (\Delta m)c^2$.

Copyright © by June Chung – All rights reserved. Reproduction of this publication without the written permission of the copyright owner is strictly prohibited.

ANSWER KEY TO EXAMPLES

Example 9-1	2.00 seconds is the time interval measured by the observer on Earth, or Δt in Eq. (9.1). The speed of the spaceship is 2.00×10^8 m/s, which is v in Eq. (9.1). Thus, $$2.00=\frac{\Delta t_o}{\sqrt{1-\left(\frac{2.00\times10^8}{3.00\times10^8}\right)^2}}$$ Solving for Δt_o, we find that $$\Delta t_o = 1.49 \text{ s}$$ The clock thus shows 11 h 32 m 01.49 s.
Example 9-2	The length of the ship is shorter when measured by the observer on Earth than when measured by the observer aboard the ship because the ship is moving relative to the observer on Earth. Use Eq. (9.4) to obtain L_o: $$25.5=L_o\sqrt{1-\frac{(6.00\times10^7)^2}{(3.00\times10^8)^2}}$$ $$L_o = 26.0 \text{ m}$$ Thus, the length of the ship is 26.0 m when the observer aboard the ship measures it. To the observer aboard the ship, the length of the meter stick on Earth will shorter than one meter because the stick is moving relative to the ship. Again, use Eq. (9.4), but this time we need to find L: $$L=1.00\sqrt{1-\frac{(6.00\times10^7)^2}{(3.00\times10^8)^2}}=0.980 \text{ m}$$ The meter stick is thus only 0.980 m long to the observer aboard the ship.
Example 9-3	We want to find the rest mass of the spaceship. Use Eq. (9.5) to find it: $$3.70\times10^5=\frac{m_o}{\sqrt{1-0.5^2}}$$ $$m_o = 3.25\times10^5 \text{ kg}$$

Copyright © by June Chung - All rights reserved. Reproduction of this publication without the written permission of the copyright owner is strictly prohibited.

PROBLEMS ON CHAPTER 9: RELATIVITY

(Answer Key on p. 236)

1 - 5. Fill in the blanks

1. When an observer is either at rest or moving at constant velocity, he or she is said to be in an _________ frame of reference.
2. According to Einstein's special theory of relativity, the speed of _________ is the value that is absolute.
3. _______ _______ is the phenomenon in which time appears to run slowly inside a fast moving object.
4. When we observe a fast moving object, its length appears shorter in the direction of travel. This is called _______ _______.
5. According to $\Delta E = (\Delta m)c^2$, _________ is a form of energy.

6 - 14. True/False questions

6. _____ No one inertial frame of reference is more special than others.
7. _____ Basic laws of physics are the same in all inertial frame of reference.
8. _____ Michelson-Morley experiment showed that light travels at different speeds depending on its direction of travel.
9. _____ The second postulate of the special theory of relativity states that, "The speed of light is the same in all inertial frame of reference."
10. _____ Two events that appear to occur simultaneously to one observer must appear so to all other observers.
11. _____ An observer aboard a spaceship can notice the watch on his wrist running more slowly as the spaceship increases its speed.
12. _____ The length of an object along the direction of travel decreases as it gains speed.
13. _____ An astronaut aboard a spaceship moving at constant velocity thinks that people on Earth have lost mass while those on Earth thinks that the astronaut has gained mass.
14. _____ At low speeds, any increase in kinetic energy results in only an increase in speed of the object. At high speeds, an increase in kinetic energy results in only an increase in mass.

Copyright © by June Chung – All rights reserved. Reproduction of this publication without the written permission of the copyright owner is strictly prohibited.

15 – 20. Multiple choice questions

15. Which of the following is not supported by Einstein's special theory of relativity?
    a) Time is not absolute.
    b) Length is not absolute.
    c) Mass is not absolute.
    d) Since time and length are not absolute, the speed of light is not absolute, either.

16. Maxwell's discovery on the speed of light was troubling to scientists at the time because
    a) the speed of light could not be confirmed through experiments
    b) the speed of light was too fast
    c) the speed of light seemed independent on the motion of the observer
    d) they did not know about the 'ether wind' then

17. Simultaneity depends on the motion of the observer because
    a) the speed of light is constant and finite
    b) the speed of light also depends on the motion of the observer
    c) different observers have different sensitivity to light
    d) some observers cannot see both events at the same time

18. An astronaut is aboard a spaceship moving at high speed. Which of the following statements is false?
    a) To observers on Earth, the astronaut appears to move slowly.
    b) To the astronaut, the observers on Earth appear to move fast.
    c) To observers on Earth, the astronaut appears thinner than usual.
    d) To the astronaut, the observers on Earth appear to have gained mass.

19. Einstein's special theory of relativity can explain which of the following phenomena?
    a) The image behind a plane mirror seems smaller in size than the actual object.
    b) An object weighs more on Earth than on the moon.
    c) The period of Earth's rotation changes.
    d) During a nuclear reaction, mass is converted to energy.

Copyright © by June Chung – All rights reserved. Reproduction of this publication without the written permission of the copyright owner is strictly prohibited.

20. It is impossible to accelerate an object initially at rest to the speed of light because

   a) it requires an infinite amount of energy

   b) time goes so slowly that it will take forever to do so

   c) its length is contracted so much that it will disappear

   d) its mass increases so much that it will create a black hole

21 - 27. Written-response questions

21. When a spaceship traveling at a speed of 1.00×10^8 m/s flies past Earth, the clock inside the spaceship shows 07 h 00 m 00.00 s. An observer on Earth turns on a red light for 2.00 seconds according to the clock on Earth. If the astronaut looks at the clock inside the spaceship after seeing the light sent from Earth, what time is the clock showing at that instant?

22. An astronaut takes a ride in a spaceship. According to the observer on Earth, the astronaut traveled for 15 days. The time of travel measured by the astronaut, on the other hand, was only 12 days. What was the speed of the spaceship?

23. A meter stick flying past Earth at an unknown speed appears to be only 0.75 m. What is the speed of the meter stick?

24. A spaceship traveling at 0.76c flies over a 100-m long field. How long does the field appear to an observer in the spaceship?

25. Observer A has twice the rest mass as observer B. When observer A, traveling past Earth at an unknown speed, looks at observer B on Earth, it appears that observer B has the same mass as herself. What is the speed of observer A?

26. During a certain nuclear chain reaction, the mass of a 2.2-kg nuclear fuel decreased by 0.01%. Find the amount of energy released during the nuclear reaction.

27. Tommy's family pays on average $72.00 on electricity per month. (The electricity company charges $0.06 per 1 kW·h.) Estimate the mass of nuclear fuel required to generate one month's supply of electric energy to Tommy's family if the conversion from mass to energy is 20% efficient.

Copyright © by June Chung – All rights reserved. Reproduction of this publication without the written permission of the copyright owner is strictly prohibited.

Copyright © by June Chung - All rights reserved. Reproduction of this publication without the written permission of the copyright owner is strictly prohibited.

CHAPTER 10

NUCLEAR PHYSICS

The cavemen knew that, if they set fire to wood, they could warm themselves and cook their food. Today we would explain this by saying that wood has chemical potential energy stored in it, and the combustion reaction releases some of this energy in the form of heat. However, the spontaneous release of energy from the nucleus of some atoms has only been known and investigated for about one hundred years, not because nuclear radiation is a new phenomenon, but because we lacked the technology to detect it until recently. Now, we are all aware of the huge energy potential hidden in the atom's nucleus. Nuclear energy has been used destructively in the atomic bomb, and constructively in nuclear reactors; and scientists have made many new elements and isotopes that benefit mankind in countless ways.

In this chapter, you will learn:

- ✓ components of the nucleus and how they are related to the atomic number and the mass number
- ✓ strong force and its role in stability of the nucleus
- ✓ binding energy
- ✓ radioactive decays: α-decay, β-decay, and γ-decay
- ✓ activity and half-life
- ✓ nuclear fission and nuclear fusion

Copyright © by June Chung – All rights reserved. Reproduction of this publication without the written permission of the copyright owner is strictly prohibited.

Key Point 10.1: **Components of the Nucleus**

discovery of sub-atomic particles

Until the late 19th century, scientists believed that atoms[†] were the smallest particles that comprise matter. However, the discovery of the electron (by J. J. Thompson in 1897) and the nucleus (by Ernest Rutherford in 1911) changed that notion, and in the early 1930s it was discovered that a nucleus consisted of protons and neutrons.

nucleons and mass number

These two constituents of a nucleus are called **nucleons**. A proton has a positive electric charge while a neutron has no electric charge. Thus, a nucleus is always electrically positive. Since the mass of the neutron is only 0.12% greater than that of the proton,[‡] their masses may be treated as equal. Thus, the mass of a nucleus is proportional to the number of nucleons, which is called the **mass number**, *A*.

atomic number

The chemical properties of a nucleus are decided by the number of protons within the nucleus. This number is called the **atomic number**, Z, and the periodic table of the elements is arranged in order of increasing atomic number, from 1 to 116. Each type of nucleus is then specified using the following notation:

$$ {}^{A}_{Z}X \qquad \ldots (10.1) $$

where *X* is the chemical symbol for the element. The number of neutrons inside a nucleus can then be found using (A – Z). For example, the nucleus of carbon 12 is denoted as ${}^{12}_{6}C$, which tells us that there are six protons and six neutrons inside the nucleus. ${}^{13}_{6}C$ is the notation for the nucleus of carbon 13, which holds six protons and seven neutrons.

isotope

Such nuclei with the same atomic number but different mass numbers are called **isotopes**.

[†] The word 'atom' comes from the Greek work 'átomos' meaning 'uncuttable.'

[‡] The mass of a proton is 1.6726×10^{-27} kg; the mass of a neutron is 1.6746×10^{-27} kg.

Copyright © by June Chung – All rights reserved. Reproduction of this publication without the written permission of the copyright owner is strictly prohibited.

- A nucleus consists of protons and neutrons. Protons and neutrons are called nucleons.
- The mass of a nucleus is proportional to the number of nucleons (or its mass number).
- The chemical properties of a nucleus are determined by the number of protons (or its atomic number).
- Each type of nucleus is specified as ${}^{A}_{Z}X$.

Example 10-1

A thorium nucleus has 234 nucleons.

(a) Find the mass number of the nucleus.

(b) Find the atomic number of the nucleus (from the periodic table).

(c) Find the number of neutrons inside the nucleus.

Copyright © by June Chung - All rights reserved. Reproduction of this publication without the written permission of the copyright owner is strictly prohibited.

Key Point 10.2: **Strong Force**

electromagnetic forces between protons

Electromagnetic forces act between all electrically charged particles. Like charges repel each other, and unlike ones attract. Neutrons, being electrically neutral, do not interact electromagnetically with electric charges. Then how could a group of protons and neutrons be packed inside a tiny nucleus? It would seem impossible to hold protons closely together because of this repulsive electromagnetic force. There must be another force binding them together.

strong forces between nucleons

This force is called the **strong force**. It is an attractive force that acts between all nucleons and is much stronger than the electromagnetic force. We do not encounter the strong force in everyday life because it is a short-range force, meaning it acts only over a range of about 10^{-15} m, approximately the diameter of a nucleus. The strong force decreases exponentially with distance and becomes insignificant for a distance greater than 10^{-15} m.

neutrons' role in stability of nucleus

Neutrons play an important role in the stability of a nucleus; by exerting a strong force on its neighboring nucleons, each neutron helps to stabilize the structure of the nucleus. However, if there are too few or too many neutrons relative to the number of protons, the nucleus is unstable, and that isotope is radioactive. (For instance, ${}^{12}_{6}C$ and ${}^{13}_{6}C$ are stable while ${}^{14}_{6}C$ is radioactive.) Notice that the ${}^{1}_{1}H$ nucleus, which is only a single proton, is the only nucleus that does not contain a neutron. (Obviously, the nucleus of the hydrogen atom needs no strong force as there is no repulsive electromagnetic force.)

- The strong force is an attractive force that acts between nucleons.
- The strong forces among nucleons keep the nucleus from falling apart.
- The strong force is a short-range force.
- Neutrons play an important role in stabilizing the nucleus.

Copyright © by June Chung – All rights reserved. Reproduction of this publication without the written permission of the copyright owner is strictly prohibited.

Key Point 10.3: **Binding Energy**

binding energy

Since the constituent particles of the nucleus are held tightly together by the strong forces amongst them, it requires energy to disassemble the nucleus. This energy is known as **binding energy**. As the number of nucleons inside the nucleus increases, so does the binding energy because each nucleon experiences the strong forces from more neighboring nucleons. However, this effect is reduced as the size of the nucleus gets larger and larger because the strong force acts over a very short range. The graph of the average binding energy per nucleon, plotted against the number of nucleons, is shown in Fig. 10-1. The average binding energy indicates the degree of bonding for each nucleon. In other words, it measures how tightly a given nucleus is held together.

average binding energy

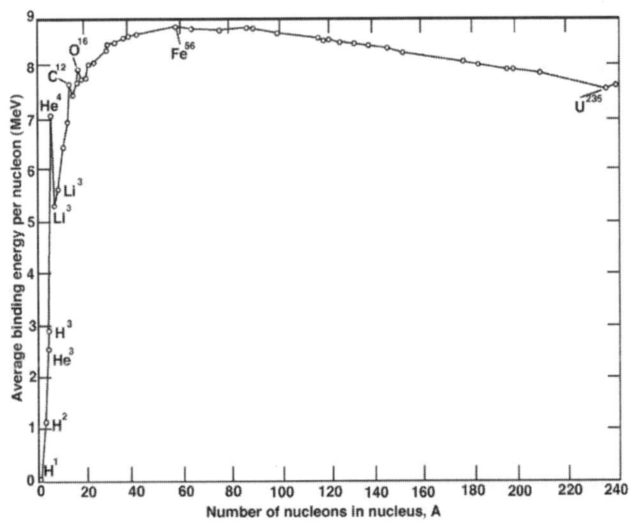

Figure 10-1: Average binding energy per nucleon vs. number of nucleons

$E = mc^2$ and binding energy

Since the binding energy is the energy we need to put into a nucleus to disassemble it, it is also the energy that is released when a nucleus is formed. This energy originates from the constituent particles of the nucleus. When protons and neutrons form a nucleus, each particle converts part of its mass into energy and releases it according to $\Delta E = (\Delta m)c^2$ (Key Point 9.8, p. 193). Thus, we can expect the total mass of a nucleus to be less than the sum of the masses of its constituents, and that is indeed the case.

Copyright © by June Chung – All rights reserved. Reproduction of this publication without the written permission of the copyright owner is strictly prohibited.

- Binding energy is the energy required to disassemble a nucleus. It is also the energy released when a nucleus is formed.
- Average binding energy per nucleon measures how tightly bound a given nucleus is.
- The total mass of a nucleus is always slightly less than the sum of the masses of its constituents.

Copyright © by June Chung – All rights reserved. Reproduction of this publication without the written permission of the copyright owner is strictly prohibited.

Key Point 10.4: **Radioactive Decays**

radioactive decay

A nucleus containing too many nucleons is like a large drop of water on the verge of bursting apart. Indeed all the isotopes of all the elements having more than 82 protons are unstable. Such unstable nuclei tend to eject particles without any external stimulus in order to reach a more stable state. This phenomenon is called **radioactive decay** and was first observed by Henri Becquerel in 1896. There are three common types of radioactive decay: alpha, beta, and gamma.

alpha decay

In **alpha decay**, a nucleus with too many protons emits an alpha particle, which is a helium-4 nucleus. The resulting nucleus has two fewer protons and neutrons than the parent nucleus. For example, when a uranium-238 nucleus undergoes alpha decay, it is transformed into a thorium-234 nucleus as shown below:

$$^{238}_{92}U \rightarrow ^{4}_{2}He + ^{234}_{90}Th \qquad \ldots (10.2)$$

Note that the superscripts and subscripts on the left equal those on the right.

beta decay

Some nuclei with too many neutrons undergo **beta decay** by emitting an electron (which is the beta particle) after converting a neutron into a proton and an electron. As a result, the atomic number of the parent nucleus increases by one while the mass number is unchanged. The following nuclear equation shows beta decay of a cesium-137 nucleus:

$$^{137}_{55}Cs \rightarrow ^{0}_{-1}e + ^{137}_{56}Ba \qquad \ldots (10.3)$$

gamma decay

Finally, in **gamma decay**, a nucleus excited to a higher energy state emits a high-energy photon called a gamma ray. Gamma decay usually follows alpha or beta decay and removes excess energy released by those decays. It does not cause the transmutation of the nucleus, as shown in the following nuclear equation:

$$^{137}_{56}Ba^{*} \rightarrow ^{137}_{56}Ba + \gamma \text{ radiation} \qquad \ldots (10.4)$$

where the symbol * denotes a nucleus in an excited energy state.

Copyright © by June Chung – All rights reserved. Reproduction of this publication without the written permission of the copyright owner is strictly prohibited.

> - Unstable nuclei tend to emit particles to reach a more stable state. This phenomenon is called radioactive decay.
> - In alpha decay, a nucleus emits an alpha particle. The atomic number decreases by two and the mass number by four as a result.
> - In beta decay, a nucleus emits a beta particle. Only the atomic number of the parent nucleus increases by one.
> - In gamma decay, a nucleus emits a gamma ray. Gamma decay does not cause nuclear transmutation.

Example 10-2

(a) Write the nuclear equation for the radium-226 nucleus undergoing alpha decay.

(b) Write the nuclear equation for the carbon-14 nucleus undergoing beta decay.

Copyright © by June Chung – All rights reserved. Reproduction of this publication without the written permission of the copyright owner is strictly prohibited.

Key Point 10.5: **Activity (a) and Half-Life (τ)**

activity

Radioactive decays occur constantly in a radioactive material. The number of nuclei that undergo radioactive decay per second is called the **activity** of the material. The activity of a sample is proportional to the number of radioactive nuclei present in the sample (N) and is expressed as

$$a = \frac{\Delta N}{\Delta t} = -\lambda N \text{ [Bq]} \qquad \dots (10.5)$$

where ΔN is the number of nuclei that undergo radioactive decay in a given time (Δt) and λ is the **decay constant**. Its unit is the Becquerel [Bq], named after Henri Becquerel.

exponential decay of activity over time

According to Eq. (10.5), the activity of a sample decreases with time as fewer and fewer radioactive nuclei remain. The resulting graph of the number of radioactive nuclei (N) versus time is shown in Fig. 10-2. Notice from the figure that it takes the same amount of time for the number of radioactive nuclei to be halved (from N_o to $N_o/2$, from $N_o/2$ to $N_o/4$, from $N_o/4$ to $N_o/8$, and so on).

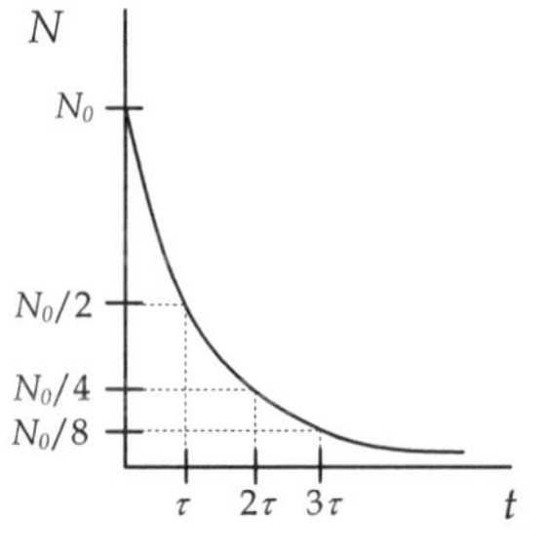

Figure 10-2: The graph of the number of radioactive nuclei vs. time

half-life

The time needed for half the remaining radioactive nuclei to undergo radioactive decay, τ, is called the **half-life** of the isotope, and has a constant value unique to a given material. A material decaying slowly has a low activity and a long half-life, and one that decays quickly has a high activity and a short half-life. The values of τ for known radioactive substances range from about 10^{-22} s to about 10^{21} years.

Copyright © by June Chung – All rights reserved. Reproduction of this publication without the written permission of the copyright owner is strictly prohibited.

> - The activity of a sample refers to the number of nuclei that undergo radioactive decay per second.
> - The activity is proportional to the number of radioactive nuclei present in a sample.
> - The number of radioactive nuclei decreases exponentially with respect to time.
> - Half-life is the time needed for half the remaining nuclei to undergo radioactive decay.

Example 10-3

The half-life of an unknown substance is 7 days. Initially, the sample contains 5.6×10^{20} radioactive nuclei. How many radioactive nuclei remain in the sample after 21 days?

Copyright © by June Chung – All rights reserved. Reproduction of this publication without the written permission of the copyright owner is strictly prohibited.

Key Point 10.6: Nuclear Fission

artificial radioactivity

In nature, radioactive decays occur spontaneously without any external stimulus. This phenomenon is called **natural radioactivity**. Unstable nuclei can also be made to undergo nuclear reaction when subjected to external stimuli. Such radioactivity is called **artificial radioactivity**. During fission, neutrons are made to collide with unstable nuclei in order to split each nucleus into two roughly equal pieces. Neutrons are very effective for this purpose because they are relatively heavy and electrically neutral.

nuclear fission

This reaction, by which an unstable nucleus is split into two pieces, is called **nuclear fission**. For example, when a uranium-235 nucleus is struck by a neutron, it can split into a krypton-92 nucleus and a barium-141 nucleus, as shown in the following nuclear equation:

$$^{235}_{92}U + {}^{1}_{0}n \rightarrow {}^{236}_{92}U \rightarrow {}^{92}_{36}Kr + {}^{141}_{56}Ba + 3\,{}^{1}_{0}n + \text{energy} \qquad \ldots (10.6)$$

During this nuclear fission, three neutrons are released at high speed, and they can be used to trigger nuclear fission in neighboring nuclei and initiate a **nuclear chain reaction**. However, because the chances of a neutron striking a particular nucleus are very small, a chain reaction will not be self-sustaining unless a sufficient amount of radioactive material is used. The smallest mass needed for a self-sustainable chain reaction is called the **critical mass**.

generation of electricity by nuclear fission

In nuclear power plants, energy extracted by nuclear fission is used to generate electricity. The amount of energy released from a chain reaction can be overwhelming, as the energy released by each nucleus quickly adds up. So, chain reactions must be kept under control for safety, and nuclear reactors contain **moderators** and **control rods** to limit the number of neutrons involved in the chain reaction and hence the rate at which energy is released.

- During a nuclear fission, an unstable nucleus is struck by a neutron and splits into two pieces.
- An enormous amount of energy can be released during a chain reaction. Nuclear plants use this energy to generate electricity.

Copyright © by June Chung – All rights reserved. Reproduction of this publication without the written permission of the copyright owner is strictly prohibited.

Key Point 10.7: **Nuclear Fusion**

nuclear fusion

While a large unstable nucleus can be made to split into two pieces, two small nuclei can collide at high speed to form one larger nucleus. This process is called **nuclear fusion**, and it is the method by which the Sun and other stars produce energy (and heavier atoms). Inside the Sun, the following nuclear fusion occurs:

$$ {}^{2}_{1}H + {}^{3}_{1}H \rightarrow {}^{4}_{2}He + {}^{1}_{0}n + \text{energy} \qquad \ldots (10.7)$$

At first, it seems amazing that a completely new nucleus could be formed by colliding two nuclei together. After all, these nuclei repel each other since both are electrically positive. Indeed, at low speeds, hydrogen nuclei will not even touch each other because of electric repulsion. However, if their speeds are great enough to bring them to within about 10^{-15} m of each other, the strong force between them will bind the nuclei together. To accelerate the nuclei to such high speeds, we need to supply enough energy to raise their temperature to about 300 million degrees Celsius, which is the temperature of the Sun's core.

nuclear fusion as energy source

Energy release during nuclear fusion is much greater than energy release during nuclear fission. We have plenty of hydrogen on Earth (about 70% of the planet is covered by water) but sustaining the high temperature and pressure environment needed for a controlled fusion reaction has proved to be extremely challenging. To date, scientists can only create the nuclear fusion of hydrogen nuclei by raising their temperature through the use of uncontrolled nuclear chain reactions.

- During a nuclear fusion, two small nuclei collide at high speed to form a larger nucleus.
- The speed of colliding nuclei must be such that they can get close enough for the strong force to bind them together.
- Energy release during nuclear fusion is much greater than that during nuclear fission.

Copyright © by June Chung – All rights reserved. Reproduction of this publication without the written permission of the copyright owner is strictly prohibited.

ANSWER KEY TO EXAMPLES

Example 10-1	a) The mass number is the number of nucleons. Thus it is 234. b) The atomic number, from the periodic table, is 90. c) The number of neutrons inside the nucleus is 234 - 90 = 144.
Example 10-2	(a) ${}^{226}_{88}Ra \rightarrow {}^{4}_{2}He + {}^{222}_{86}Rn$ (b) ${}^{14}_{6}C \rightarrow {}^{0}_{-1}e + {}^{14}_{7}N$
Example 10-3	After 7 days, 2.8×10^{20} nuclei remain. After 14 days, half of that remain. Finally, after 21 days, 0.7×10^{20} nuclei remain in the sample.

Copyright © by June Chung - All rights reserved. Reproduction of this publication without the written permission of the copyright owner is strictly prohibited.

PROBLEMS ON CHAPTER 10: NUCLEAR PHYSICS

(Answer Key on p. 237)

1 – 10. Fill in the blanks

1. The _______ _______ of a nucleus is determined by the number of nucleons inside it.
2. __________ force is an attractive force that acts between all nucleons.
3. The __________ energy of a nucleus is the energy required to disassemble it.
4. In __________ decay, a nucleus with too many protons emits a helium-4 nucleus.
5. In __________ decay, an electron, generated when a neutron is converted into a proton and an electron, is emitted.
6. The number of nuclei that undergo radioactive decays per second is the __________ of the material.
7. The time needed for half the remaining radioactive nuclei to undergo radioactive decay is called _______ _______.
8. In a nuclear __________, an unstable nucleus is split into two smaller nuclei.
9. The smallest mass for a self-sustainable chain reaction is called the __________ mass.
10. In a nuclear __________, two small nuclei collide at high speed to form one larger nucleus.

11 – 20. True/False questions

11. _____ Protons, neutrons, and electrons are called nucleons.
12. _____ Atomic number is determined by the number of nucleons inside a nucleus.
13. _____ The weak force is an attractive force that acts between all nucleons.
14. _____ Binding energy is the energy required to disassemble a nucleus.
15. _____ An electron is emitted during an alpha decay.
16. _____ The activity of a sample decreases linearly with time.
17. _____ A material decaying slowly has a low activity and a long half-life.

Copyright © by June Chung – All rights reserved. Reproduction of this publication without the written permission of the copyright owner is strictly prohibited.

18. _____ Usually electrons are used to trigger nuclear fission reactions because they are attracted electrically by unstable nuclei.

19. _____ The strong force between two colliding nuclei enable nuclear fusion reactions.

20. _____ Nuclear plants use nuclear fusion reactions to generate electricity.

21 - 25. Multiple choice questions

21. The total mass of a nucleus is always slightly less than the sum of the masses of its constituents because

    a) each nucleon converts part of its mass into binding energy
    b) the strong force packs the nucleons tightly
    c) each nucleon becomes smaller when inside a nucleus
    d) protons are attracted by the surrounding electrons when inside a nucleus

22. Which of the following statements is <u>not</u> true for beta decay?

    a) One of the neutrons inside a nucleus is converted into a proton and an electron.
    b) An electron is emitted from the nucleus.
    c) The atomic number remains unchanged.
    d) The mass number remains unchanged.

23. What is the purpose of control rod?

    a) to absorb neutrons
    b) to slow down neutrons
    c) to absorb energy released from fission reactions
    d) all of the above

24. Which of the following statements is <u>not</u> true for fission and fusion reactions?

    a) During a nuclear fission, there is a small loss in total mass.
    b) During a nuclear fusion, there is a small gain in total mass.
    c) During a nuclear fission, an unstable nucleus is split into two pieces.
    d) During a nuclear fusion, two nuclei are joined together to form a larger nucleus.

Copyright © by June Chung - All rights reserved. Reproduction of this publication without the written permission of the copyright owner is strictly prohibited.

25. Stars generate energy mainly through __________.

    a) chemical reactions
    b) gamma radiations
    c) nuclear fission
    d) nuclear fusion

26 - 30. Written-response questions

26. A plutonium nucleus has 239 nucleons. Find the number of neutrons inside the nucleus.

27. Using Fig. 10-1, estimate the total mass of the deuterium nucleus, ${}_{1}^{2}H$. (Use the following conversion factor: 1 eV = 1.6×10^{-19} J)

28. Write the nuclear equation for an Einsteinium-254 nucleus undergoing alpha decay and gamma decay in succession.

29. An unknown nucleus is struck by a neutron and undergoes a nuclear reaction written as

$${}_{A}^{Z}X+{}_{0}^{1}n\rightarrow{}_{79}^{197}Au+{}_{1}^{2}H$$

Determine the identity of the unknown nucleus, ${}_{A}^{Z}X$.

30. The half-life of an unknown substance is 75 seconds. If the sample contains 3.9×10^{17} radioactive nuclei, how many radioactive nuclei remain after five minutes?

Copyright © by June Chung - All rights reserved. Reproduction of this publication without the written permission of the copyright owner is strictly prohibited.

APPENDIX A. CHECK LIST

Chapter 1. Introduction – You should be able to

- ☐ Perform mathematical operations using significant digits.
- ☐ Define vector and scalar quantities and distinguish between them.
- ☐ Perform vector additions and subtractions.

Chapter 2. Kinematics – You should be able to

- ☐ Define position, displacement, distance, velocity, speed, and acceleration and understand how they are related.
- ☐ Determine velocity and speed from a position vs. time graph and displacement and acceleration from a velocity vs. time graph.
- ☐ Represent various motions in position vs. time and velocity vs. time graphs.
- ☐ State the characteristics of free-falling objects.
- ☐ Solve problems using kinematic equations.

> Make sure the object has a constant acceleration!

> If the use of kinematic equations does not work, try solving the problem using the **work-energy principle**!

Chapter 3. Dynamics – You should be able to

- ☐ State Newton's laws of motion and provide example situations involving each law.
- ☐ State differences between mass and weight.
- ☐ Explain how to obtain the magnitude and direction of various forces ($\vec{F}_g$, $\vec{F}_N$, $\vec{F}_T$, $\vec{F}_f$).
- ☐ Draw free-body diagrams of objects in various situations.
- ☐ Solve dynamic problems.

> Include only the forces 'on' the object!

- Take these three steps…
  1. Draw the FBD of the object of interest.
  2. Set the positive direction, and determine the sign of each force.
  3. Apply $\vec{F}_{net} = m\vec{a}$ in each direction and solve for the unknown.

> If there are two masses connected by a string, draw the free-body diagram of each mass. After applying $\vec{F}_{net} = m\vec{a}$ to each mass, you should have two equations and two unknowns!

Copyright © by June Chung – All rights reserved. Reproduction of this publication without the written permission of the copyright owner is strictly prohibited.

Chapter 4. Momentum – You should be able to

- ☐ Define momentum, change of momentum, and impulse, and describe how they are related.
- ☐ State the law of conservation of momentum, and explain the condition for it.
- ☐ Solve collision and explosion problems using conservation of momentum.

> Remember that the equation for conservation of momentum is a <u>vector</u> equation!

Chapter 5. Work, Energy, and Power – You should be able to

- ☐ Define work and energy, and describe how they are related.
- ☐ State the conservation of mechanical energy and explain the condition for it.
- ☐ Solve work-energy problems using the conservation of mechanical energy.
- ☐ Define power and efficiency.

> There should not be a friction or applied force!

Chapter 6. Thermal Physics – You should be able to

- ☐ Define temperature, thermal energy, and heat, and describe how they are different.
- ☐ Define specific heat capacity and latent heat, and calculate the amount of thermal energy released from or absorbed by an object during heating or cooling.
- ☐ State the conservation of heat for a system, and use it to solve problems involving heat flow within a system.

Chapter 7. Waves – You should be able to

- ☐ Define wave motion and describe transverse and longitudinal waves, giving examples.
- ☐ Use the wave equation to determine the frequency and the wavelength of a wave.
- ☐ Describe the following properties of waves: reflection, refraction, diffraction, interference, and the Doppler effect
- ☐ Explain when and how total internal reflection occurs.
- ☐ Determine the resulting displacement of two interfering waves based on the principle of superposition.
- ☐ Describe how standing waves occur, and determine the frequency of the n-th mode using the wave equation.

Copyright © by June Chung – All rights reserved. Reproduction of this publication without the written permission of the copyright owner is strictly prohibited.

Chapter 8. Geometric Optics – You should be able to

Real or virtual?
Upright or inverted?
Size compared to object?

- ☐ Describe how our eyes see objects and their images.
- ☐ Describe image formation by a plane mirror and the characteristics of such an image.
- ☐ Construct ray diagrams for spherical mirrors, and describe the characteristics of images formed by concave and convex mirrors.
- ☐ Construct ray diagrams for thin lenses (bi-convex and bi-concave), and describe the characteristics of those images.
- ☐ Use mirror/lens equation to find the object/image distance and the height of object/image.

Watch out for the sign convention of the image distance and height!

Chapter 9. Relativity – You should be able to

- ☐ Describe why Maxwell's prediction on the speed of light was troubling to scientists at the time.
- ☐ List the postulates of Einstein's special theory of relativity.
- ☐ Describe how two events that appear to occur simultaneously to one observer may not appear so to another.
- ☐ Describe time dilation, length contraction, and mass increase.
- ☐ Explain the significance of Einstein's energy-mass equation: $E = mc^2$

Chapter 10. Nuclear Physics – You should be able to

- ☐ Define atomic number and mass number, and explain how they are related to the constituents of the nucleus.
- ☐ Describe the strong force and how it helps stabilize a nucleus.
- ☐ Explain the role of the neutron in the stability of a nucleus.
- ☐ Define binding energy, and explain what binding energy per nucleon represents.
- ☐ Explain why a nucleus undergoes radioactive decay, and describe α-, β-, and γ-decays.
- ☐ Describe the nuclear fission and fusion reactions.

Copyright © by June Chung - All rights reserved. Reproduction of this publication without the written permission of the copyright owner is strictly prohibited.

APPENDIX B. UNITS

The fundamental units in physics are kg, m, s, and K, and all other units can be expressed in terms of those fundamental units. In the table below, the physical quantities introduced in this course are shown with their standard units.

Quantity	Standard unit	Alternate expression
Mass	kg	
Distance, Displacement	m	
Time	s	
Speed, Velocity	m/s	
Acceleration	m/s^2	
Force	N	$N = kg{\cdot}m/s^2$
Momentum, Impulse	N·s	N·s = kg·m/s
Work, Heat, Energy(potential/kinetic/thermal)	J	$J = N{\cdot}m = W{\cdot}s = kg{\cdot}m^2/s^2$
Power	W	$W = J/s = kg{\cdot}m^2/s^3$
Temperature	K	K = °C + 273.15
Specific heat	J/(kg °C)	
Latent heat	J/kg	
Frequency	Hz	$Hz = s^{-1}$
Period	s	
Wavelength	m	
Activity	Bq	
Half-life	s	

Copyright © by June Chung – All rights reserved. Reproduction of this publication without the written permission of the copyright owner is strictly prohibited.

APPENDIX C. EQUATIONS

Chapter 2: Kinematics

Displacement: $\vec{d} = \Delta\vec{x}$

Velocity: $\vec{v} = \dfrac{\Delta\vec{x}}{\Delta t} = \dfrac{\vec{d}}{\Delta t}$ Speed: $v = \dfrac{\text{distance}}{\text{time taken}} = \dfrac{d}{\Delta t}$

Acceleration: $\vec{a} = \dfrac{\Delta\vec{v}}{\Delta t}$

Kinematic equations: If $\vec{a}$ is constant,

$$\vec{v}_{avg} = \frac{\vec{v}_i + \vec{v}_f}{2} \qquad \vec{d} = \vec{v}_{avg}\, t$$

$$\vec{v}_f = \vec{v}_i + \vec{a}t \qquad \vec{d} = \vec{v}_i t + \frac{1}{2}\vec{a}t^2$$

$$v_f^2 = v_i^2 + 2\vec{a}\cdot\vec{d}$$

Chapter 3: Dynamics

Newton's first law of motion: If $\vec{F}_{net} = 0$, then $\vec{v} = \text{constant}$.

Newton's second law of motion: $\vec{F}_{net} = m\,\vec{a}$

Newton's third law of motion: $\vec{F}_{on\,A\,by\,B} = -\vec{F}_{on\,B\,by\,A}$

Force of gravity: $\vec{F}_g = m\,\vec{g}$

Force of friction: $F_f = \mu F_N$

Chapter 4: Momentum

Momentum: $\vec{p} = m\,\vec{v}$

Change in momentum: $\Delta\vec{p} = m\left(\vec{v}_f - \vec{v}_i\right)$

Impulse: $\vec{J} = \vec{F}_{net}\,\Delta t = \Delta\vec{p} = m\left(\vec{v}_f - \vec{v}_i\right)$

Conservation of momentum: $\left(m_1\vec{v}_1 + m_2\vec{v}_2\right)_{before} = \left(m_1\vec{v}_1 + m_2\vec{v}_2\right)_{after}$

Conservation of kinetic energy for a perfectly elastic collision:

$$\left(\frac{1}{2}m_1 v_1^2 + \frac{1}{2}m_2 v_2^2\right)_{before} = \left(\frac{1}{2}m_1 v_1^2 + \frac{1}{2}m_2 v_2^2\right)_{after}$$

Copyright © by June Chung – All rights reserved. Reproduction of this publication without the written permission of the copyright owner is strictly prohibited.

Chapter 5: Work, Energy, and Power

Work: $W = F \cdot d$ if force and displacement are in the same direction

$W = -F \cdot d$ if force and displacement are in opposite directions

$W = 0$ if force and displacement are at a right angle

Kinetic energy: $E_K = \frac{1}{2} m v^2$

Potential energy: $E_P = m g h$

Work-energy principle: $W = \Delta E_K$

Power: $P = \frac{W}{t} = F \cdot v$

Efficiency: $Efficiency = \frac{P_{out}}{P_{in}} = \frac{W_{out}}{W_{in}}$

Chapter 6: Thermal Physics

Heat into or out of a mass: $Q = m c \Delta T$ if there is no change of phase

$Q = m c \Delta T + mL$ if the mass changes phase

Conservation of heat: $Q_{lost\ by\ A} = Q_{gained\ by\ B}$

Chapter 7: Wave

Wave equation: $v = \lambda f = \frac{\lambda}{T}$

Law of reflection: $\theta_i = \theta_r$

Law of refraction: $\frac{\sin \theta_i}{\sin \theta_r} = \frac{v_i}{v_r}$

Frequency of n-th mode in a standing wave: $f_n = n f_1$

Copyright © by June Chung – All rights reserved. Reproduction of this publication without the written permission of the copyright owner is strictly prohibited.

Chapter 8: Geometric Optics

Mirror/Lens equation: $\frac{1}{d_o}+\frac{1}{d_i}=\frac{1}{f}$

Magnification factor: $m=\frac{h_i}{h_o}=-\frac{d_i}{d_o}$

Chapter 9: Relativity

Time dilation: $\Delta t=\frac{\Delta t_o}{\sqrt{1-\frac{v^2}{c^2}}}$

Length contraction: $L=L_o\sqrt{1-\frac{v^2}{c^2}}$

Mass increase: $m=\frac{m_o}{\sqrt{1-\frac{v^2}{c^2}}}$

Energy-mass equation: $E=mc^2=m_oc^2+E_K$

Chapter 10: Nuclear Physics

Alpha decay: ${}^{A}_{Z}X \rightarrow {}^{4}_{2}He+{}^{A-4}_{Z-2}Y$

Beta decay: ${}^{A}_{Z}X \rightarrow {}^{0}_{-1}e+{}^{A}_{Z+1}Y$

Gamma decay: ${}^{A}_{Z}X^{*} \rightarrow {}^{A}_{Z}X+\gamma$ radiation (* denotes an excited energy state)

Activity: $a=\frac{\Delta N}{\Delta t}=-\lambda N$

Copyright © by June Chung – All rights reserved. Reproduction of this publication without the written permission of the copyright owner is strictly prohibited.

APPENDIX D. PERIODIC TABLE

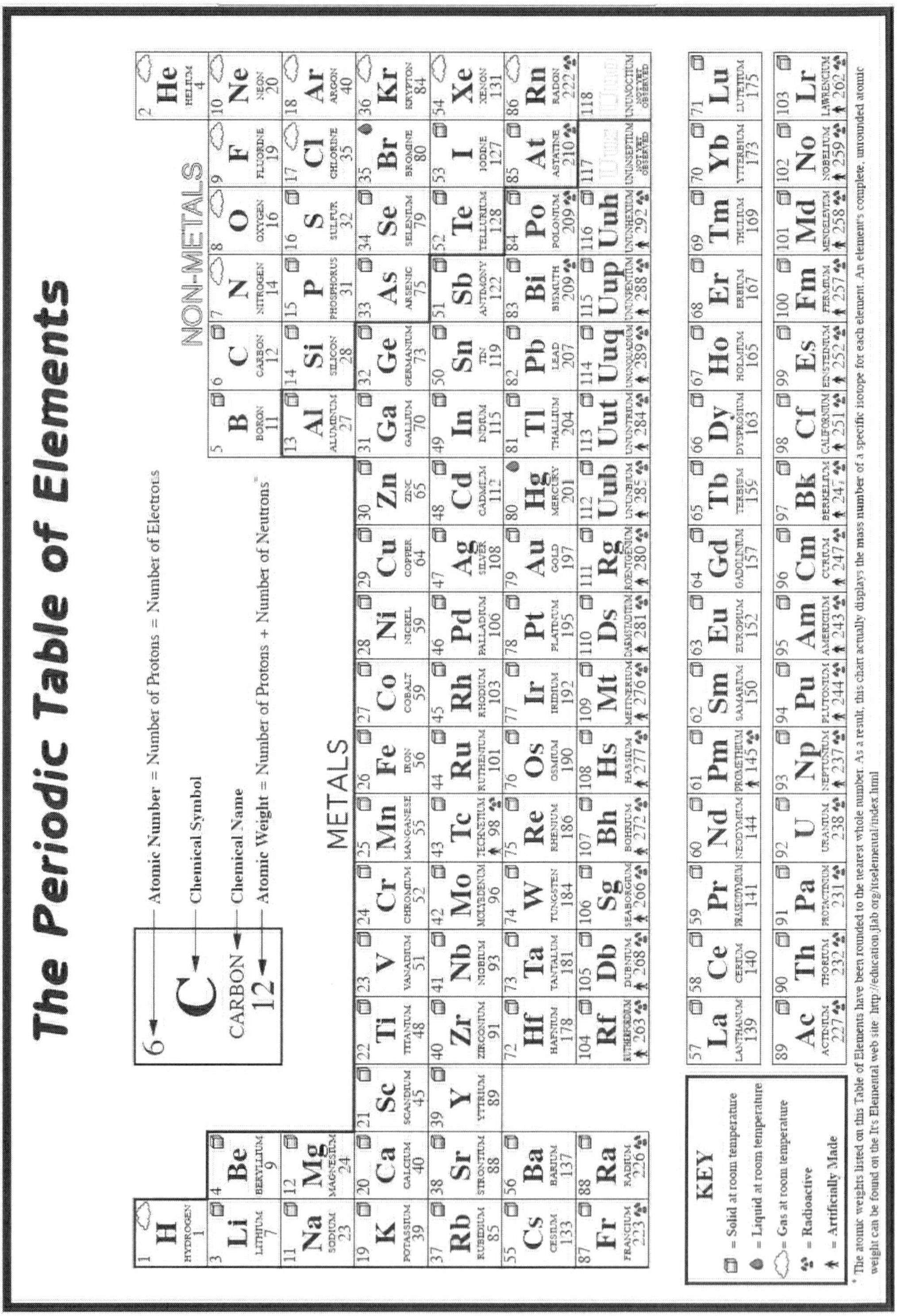

Copyright © by June Chung – All rights reserved. Reproduction of this publication without the written permission of the copyright owner is strictly prohibited.

APPENDIX E. ANSWER KEY

CHAPTER 1: INTRODUCTION

1. significant digits	2. kg, m, s	3. metric prefixes
4. scalar	5. vector	

6. False – 2.100 has four significant digits.
7. True
8. True
9. True
10. True
11. False – Two parallel vectors may point in opposite directions.
12. True
13. True
14. True
15. True

16. b	17. d	18. c	19. c	20. d

21. a) 15.482 cm = 15.48 cm (up to the second decimal digit) = 1.548×10^1 cm
    b) 8.4581 s = 8.458 s (up to the third decimal digit)
    c) 374.2675 m^2 = 374 m^2 (keep only three significant digits) = 3.74×10^2 m^2
    d) 173.9726 km/hr = 170 km/hr (keep only two significant digits) = 1.7×10^2 km/hr
22. a) 3510 MHz
    b) 290 ns
    c) 0.670 MByte
    d) 0.0000048 mg
23. $2\vec{A} = 7.4$ m, north, and $3\vec{B} = 4.5$ m, south. $-3\vec{B} = 4.5$ m, north. Then $2\vec{A} - 3\vec{B} = 11.9$ m, north
24. 1.9 km, south
25. The two vectors and their resultant vector form a right triangle as shown. Use the Pythagorean theorem to find the length of the resultant vector, which is 13.0 m.

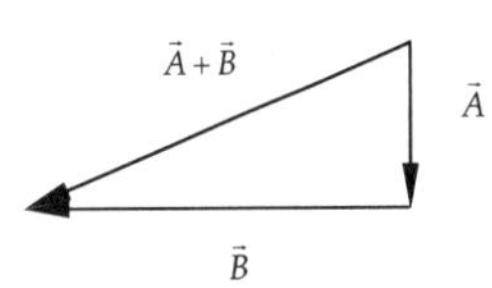

CHAPTER 2: KINEMATICS

1. position	2. displacement	3. distance
4. velocity	5. speed	6. acceleration
7. displacement	8. velocity	9. acceleration
10. free-falling		

Copyright © by June Chung – All rights reserved. Reproduction of this publication without the written permission of the copyright owner is strictly prohibited.

11. True
12. True
13. False – The displacement of a moving body could be zero if its initial and final positions are the same.
14. True
15. False – If the velocity is constant, the speed must also be constant.
16. False – The direction of acceleration is independent of that of velocity.
17. False – The area under the velocity vs. time curve represents displacement.
18. True
19. True
20. False – Both objects hit the ground simultaneously.

21. b	22. d	23. d	24. b	25. c	26. b
27. c	28. c				

29. Let the distance from A to B be d. Then the total distance of travel is $2d$. The total time of travel is

$$t = \frac{d}{2.5} + \frac{d}{2.0} = 0.90d$$

Therefore, the average speed for the entire course is 2.2 m/s.

30. a) The area under the curve for A = 10 t_1 m, and that for B is 10.5 t_1 m. Thus, B travels farther in the positive direction than A.

b) Since the average velocity is the displacement over time, it is 10 m/s for A and 10.5 m/s for B. Thus, B travels at the greater speed in the positive direction than A.

c) At $t = t_1$, A has the velocity of +20 m/s while B has the velocity of + 16 m/s. Thus, A moves at the greater speed in the positive direction than B at $t = t_1$.

d) The slope of the curve for A is $20/t_1$ m/s², and that for B is $11/t_1$ m/s². Thus, A accelerates more in the positive direction than B.

31. $d = \frac{1}{2}(2.50)(6.67)^2 + (16.7)(120) = 2060 \text{ m}$

32.

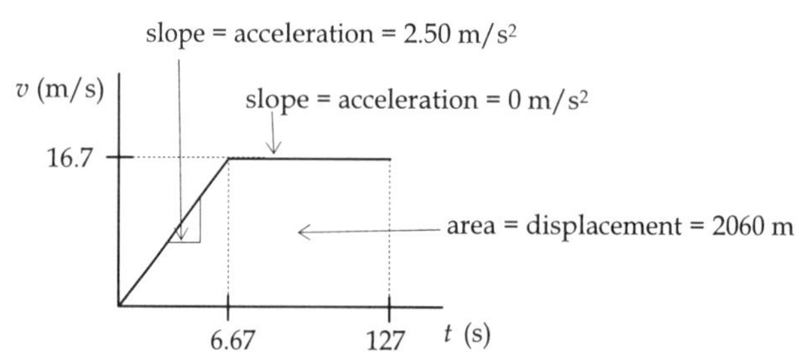

33. a) $a = \frac{v_f - v_i}{t} = \frac{-35-15}{5.0} = -10 \text{ m/s}^2$; thus, the object accelerated at 10 m/s² to the left.

b) $v_f = 0 = v_i + at = 15 + (-10)t$; it took 1.5 seconds for the object to come to rest momentarily.

c) $d = v_i t + \frac{1}{2}at^2 = 15(5.0) + \frac{1}{2}(-10)(5.0)^2 = -50$ m; its displacement was 50 m to the left.

During the first 1.5 seconds, it traveled 11.25 m to the right. During the next 3.5 seconds, it traveled 61.25 m to the left. Thus, the total distance of travel is 72.5 m.

Copyright © by June Chung – All rights reserved. Reproduction of this publication without the written permission of the copyright owner is strictly prohibited.

34. The distance traveled by the first object: $d = \frac{1}{2}(3.0)(2.4)^2 = 8.64$ m.

The minimum acceleration required for the second object: $8.64 = 5.3(1.2) + \frac{1}{2}a(1.2)^2$; $a =$ 3.2 m/s^2

35. The initial and final speeds of the block are 17 m/s and 0 m/s, respectively, and the displacement is 22 m. To find the deceleration of the block, use Eq. (2.9): $0^2 = (17)^2 + 2a(22)$; $a =$ -6.6 m/s^2

Thus, the block decelerates at 6.6 m/s^2.

36.

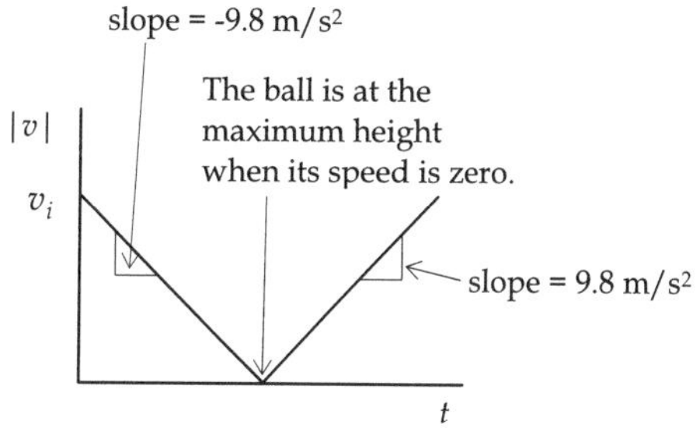

37. The velocity of the ball at the beginning of the third second, or at t = 2.0 s, is $v_2 = 22 + (-9.8)(2.0) = +2.4$ m/s. The velocity at the end of the third second, or at t = 3.0 s, is $v_3 = 22 + (-9.8)(3.0) = -7.4$ m/s. The change in the direction of velocity means that the ball reaches the maximum height during the third second of travel. Thus, we need to find the position of the ball at t = 2.0 s, t = 3.0 s, and at the maximum height:

At t = 2.0 s, $d = 22(2.0) + \frac{1}{2}(-9.8)(2.0)^2 = 24.4$ m above the ground

The maximum height: $0^2 = (22)^2 + 2(-9.8)d$; $d = 24.7$ m above the ground

At t = 3.0 s, $d = 22(3.0) + \frac{1}{2}(-9.8)(3.0)^2 = 21.9$ m above the ground

The total distance of travel during the third second is $d = (24.7 - 24.4) + (24.7 - 21.9) = 3.1$ m.

38. $v_i^2 = -2ad = -2(-9.8)(15) = 294$ m^2/s^2; $v_i = 17$ m/s

39. $d = 12(5.1) + \frac{1}{2}(-9.8)(5.1)^2 = -66$ m; Thus, the building is 66 m high.

Maximum height of the ball = 7.3 m above the building. The total distance of travel = 81 m.

Average speed = 81/5.1 = 16 m/s

Average velocity = 66/5.1 = 13 m/s, toward the ground

40. Let the speed of the object at the top of the window be v_1. Setting the downward direction as positive, we can find v_1 using Eq. (2.8): $2.4 = v_1(0.15) + \frac{1}{2}(+9.8)(0.15)^2$; $v_1 = 15.265$ m/s.

Find how far the object has fallen from the top of the building: $(15.265)^2 = 0^2 + 2(+9.8)d$; $d = 11.9$ m.

Therefore, the object was dropped 12 m above the top of the window.

Copyright © by June Chung – All rights reserved. Reproduction of this publication without the written permission of the copyright owner is strictly prohibited.

CHAPTER 3: DYNAMICS

1. force	2. inertia	3. net force
4. inertia	5. mass	6. weight
7. normal	8. friction	9. tension
10. free-body diagram		

11. False - If the forces are balanced, the object may or may not move. If it moves, its velocity is constant.
12. True
13. True
14. False - Newton's First Law states that "if the forces on a body are balanced, its acceleration is zero."
15. False - The unit for weight is the Newton [N].
16. False - The magnitude of the normal force may be different from that of the object's weight if an applied force acts on the object vertically.
17. True
18. True
19. False - An object may experience a friction force even when it is at rest.
20. False - The magnitude of friction force is independent of the speed of the object.

21. b	22. b	23. d	24. d	25. a
26. b	27. b	28. a		

29. Using Eq. (2.8), we can find the acceleration of the object: $73 = \frac{1}{2}a(5.5)^2$; $a = 4.8$ m/s²

    Thus, $F_{net} = ma = (150)(4.8) = 720$ N

30. According to Newton's Third Law, the force on the 66-kg object is $(350)(0.28) = 98$ N, opposite direction of its initial motion. Thus, its acceleration is $\frac{98}{66} = 1.5$ m/s², opposite to the direction of its initial motion.

31. $m_A = \frac{F}{7.2}$ and $m_B = \frac{F}{5.1}$.

    Therefore, $F = (m_A + m_B)\,a = \left(\frac{F}{7.2} + \frac{F}{5.1}\right)a = 0.335Fa$; $a = 3.0$ m/s²

32. The free-body diagram of the object is shown at right. As the magnitude of the tension force increases, the acceleration of the object increases as well. When the tension force becomes 55 N, the equation of motion for the object becomes

    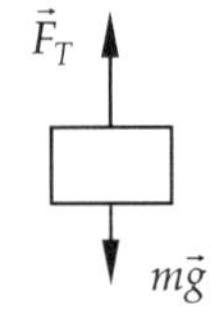

    $$55 - (2.2)(9.8) = (2.2)a$$

    The maximum acceleration the object can attain is 15 m/s².

33. Since the object remains at rest, the net force acting on the object must be zero, and the horizontal applied force must be balanced by the friction force. Thus, the magnitude of the friction force is 29 N.

34. a) To find the acceleration of the elevator, which is also Eric's acceleration, use Eq. (2.7):

    $$7.2 = 0 + a(8.5);\quad a = 0.85 \text{ m/s}^2$$

    Eric experiences two forces, the gravitational and normal force, and their sum, or the net force, is

    $$F_{net} = F_N - F_g = ma = m(0.85)$$

Copyright © by June Chung - All rights reserved. Reproduction of this publication without the written permission of the copyright owner is strictly prohibited.

Since Eric weight is 580 N, his mass is 580/9.8 = 59.2 kg. Thus, the normal force on Eric while he is accelerating upward is

$$F_N = 580 + (59.2)(0.85) = 630 \text{ N}$$

b) Because the elevator decelerates from 7.2 m/s to 0 m/s in 25 m, we can find its deceleration using Eq. (2.9):

$$0^2 = 7.2^2 + 2a(25); \ a = -1.04 \text{ m/s}^2$$

The equation of motion is thus

$$F_{net} = F_N - F_g = ma = m(-1.04)$$

This time, the magnitude of the normal force on Eric's body is less than that of his weight, and it is

$$F_N = 580 + (59.2)(-1.04) = 520 \text{ N}$$

35. The equation of motion for the 8.3-kg object: $F_T = (8.3)a$

The equation of motion for the 4.0-kg object: $(4.0)(9.8) - F_T = (4.0)a$

Solve the system of equations for the unknowns: $F_T = 3.2$ N, $a = 0.38$ m/s²

36. a) There are two forces acting on the object: the gravitational force acting downward and the upward applied force. The net force on the object is $F_A - F_g = 22.0 - (1.50)(9.8) = 7.3$ N.

b) While the object is being accelerated in Jose's hand, its acceleration is $a = \dfrac{7.3}{1.50} = 4.87$ m/s². We can find the velocity of the object as it leaves Jose's hand using Eq. (2.9):

$$v^2 = 0^2 + 2(4.87)(0.75); \ v = 2.70 \text{ m/s}$$

Using 2.7 m/s as the initial speed of the object, we can find the maximum height above Jose's hand from Eq. (2.9). Note that the acceleration of the object is now -9.8 m/s².

$$0^2 = (2.7)^2 + 2(-9.8)d; \ d = 0.372 \text{ m}$$

37. The free-body diagrams of the objects are shown below.

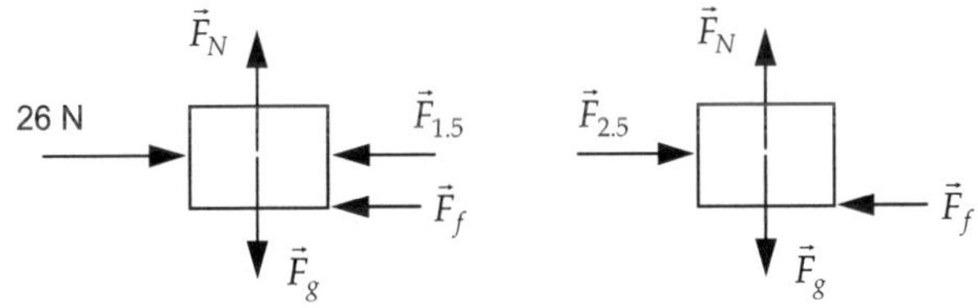

The equation of motion for the 2.5-kg object: $26 - F_{1.5} - 0.47(2.5)(9.8) = (2.5)a$

The equation of motion for the 1.5-kg object: $F_{2.5} - 0.47(1.5)(9.8) = (1.5)a$

Note that $F_{1.5} = F_{2.5}$ as they are an action-reaction pair. Solve the system of equations, and we find that

$$F_{1.5} = F_{2.5} = 17 \text{ N}$$

Thus, both objects push each other with 17-N force.

38. a) The equation of motion for the 19-kg object: $(19)(9.8) - F_T = (19)a$

The equation of motion for the 11-kg object: $F_T - (11)(9.8) = (11)a$

Solve the system of equation for the unknowns: $F_T = 140$ N, $a = 2.6$ m/s²

b) Set the positive direction as downward, and apply Eq. (2.8): $0.75 = 0 + \frac{1}{2}(2.6)t^2; \ t = 0.76$ s

Copyright © by June Chung – All rights reserved. Reproduction of this publication without the written permission of the copyright owner is strictly prohibited.

c) Use Eq. (2.9): $v_f^2 = 0^2 + 2(2.6)(0.75)$; $v_f = 2.0$ m/s

The velocity of the 11-kg object is thus 2.0 m/s, upward.

39. a) Since the car is traveling at constant velocity, the net force on the car must be zero.

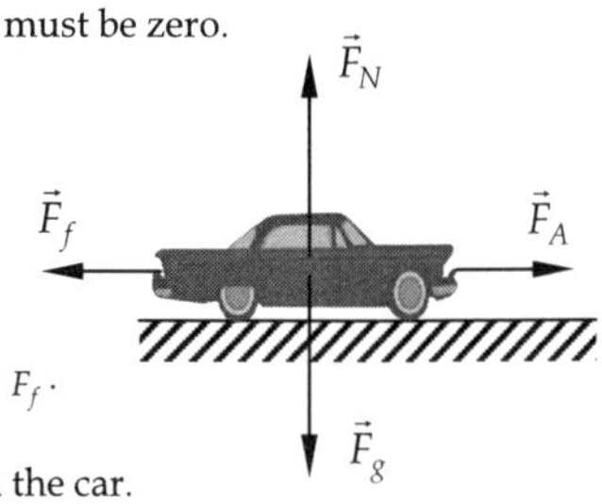

The force of gravity: $F_g = (1200)(9.8) = 12000$ N

The normal force: $F_N = 12000$ N because it must balance F_g.

The friction force: $F_f = \mu F_N = (0.27)(12000) = 3200$ N

The applied force of the engine: $F_A = 3200$ N because it must balance F_f.

b) When the car runs out of gas, the engine can no longer exert force on the car. Thus, the net force on the car is 3200 N, acting in the opposite direction of motion. The acceleration of the car is then

$$a = \frac{3200}{1200} = 2.6 \text{ m/s2, in the opposite direction of motion}$$

To determine how far the car can travel without gas, use Eq. (2.9). (The initial velocity of the car is 90 km/h = $90 \times \frac{1000}{3600} = 25$ m/s, in the direction of motion.)

$$0^2 - 25^2 + 2(-2.6)d;\ d = 118 \text{ m}$$

Therefore, the car will not reach the gas station before coming to a half.

40. The free-body diagram of the block moving with the bullet embedded is shown right. The magnitude of the block's gravitational force is $(1.525)(9.8) = 14.95$ N. The normal force, as it balanced the weight of the block, is also 14.95 N. The magnitude of the friction force on the block is then $\mu F_N = (0.45)(14.95) = 6.73$ N, which is equal to the magnitude of the net force. The acceleration of the block is thus

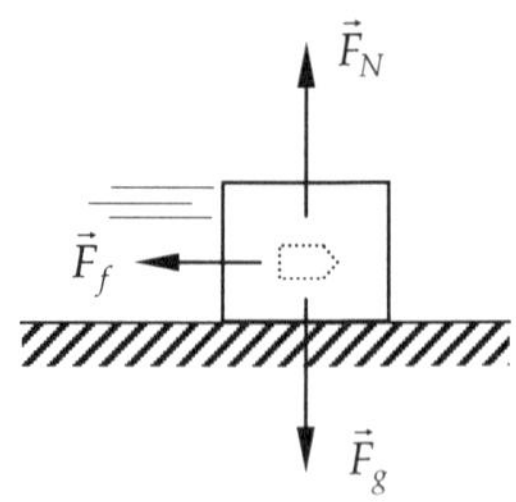

$$a = \frac{6.73}{1.525} = 4.41 \text{ m/s}^2$$

To determine how far the block travels before coming to a stop, use Eq. (2.9):

$$0^2 = 13^2 + 2(-4.41)d;\ d = 19 \text{ m}$$

41. On the surface of the moon, the magnitude of the gravitational force is reduced for both the block and the bullet. Thus, the magnitude of the normal force is reduced, which in turn reduces the magnitude of the friction force on the block. As a result, the block travels <u>farther</u> on the surface of the moon.

42. The free-body diagram of the bullet is shown. Note that we ignored the weight of the bullet. The acceleration of the bullet is found from Eq. (2.9):

$$0^2 = 375^2 + 2a(0.14);\ a = -5.0 \times 10^5 \text{ m/s}^2$$

Thus, the net force on the bullet is

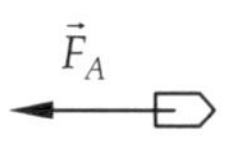

$$F_{net} = ma = (0.035)(-5.0 \times 10^5) = 1.8 \times 10^4 \text{ N}$$

which is the average force exerted by the wall. The force opposes the bullet's motion.

Copyright © by June Chung - All rights reserved. Reproduction of this publication without the written permission of the copyright owner is strictly prohibited.

CHAPTER 4: MOMENTUM

1. momentum, velocity	2. net force	3. impulse, momentum
4. perfectly inelastic		

5. True
6. False – The greater the change in momentum, the greater the net force.
7. True
8. True
9. True
10. True
11. False – The sum of colliding object's momentum is conserved in all collisions.
12. True

13. d	14. b	15. a	16. b	17. c	18. d

19. Use Eq. (2.7) to find the velocity of the object after 1.8 s: $v_f = 14 + (-9.8)(1.8) = -3.64$ m/s

    Since $F_g = mg = m(9.8) = 28$ N, the object's mass is 2.86 kg. Thus, the object's momentum is

    $\vec{p} = m\vec{v} = 2.86(-3.64) = -10.4$ kg·m/s, or 10 kg·m/s, vertically downward.

20. $p_f = (15)v_f = F\Delta t + mv_i = 30(3.0) + 0 = 90$ kg·m/s; $v_f = 6.0$ m/s

21. The change in momentum of the object is $\Delta\vec{p} = 7.3(0-16) = -117$ kg·m/s. If this change in momentum occurs in 0.75 seconds, the average net force on the object is $\vec{F}_{net} = \dfrac{\Delta\vec{p}}{\Delta t} = \dfrac{-117}{0.75} = -156$ N. Thus, the minimum average net force on the object is 156 N, in the opposite direction of its motion.

22. Object B needs half the impulse given to object A. Impulse given to A is 125 N s. Thus, we must apply the 10-N force for 6.3 seconds.

23. Use Eq. (2.8) to find the acceleration of the object: $5.5 = 0 + \frac{1}{2}a(2.0)^2 = 2.75$ m/s²

    The velocity of the object after 2.0 seconds is then $v_f^2 = 0^2 + 2(2.75)(5.5) = 5.5$ m/s.

    The change in momentum of the object is $\Delta\vec{p} = 9.5(5.5-0) = 52.3$ kg·m/s.

    Thus, the impulse on the object is 52 kg·m/s, in the direction of the applied force.

24. $F\Delta t = (-54)(0.33) = m(v_f - v_i) = m(1.8 - 2.1)$; $m = 59$ kg

25. $\Delta p = 0.110(-13.8 - 14..2) = -3.08$ N s; $F_{avg} = \dfrac{\Delta p}{\Delta t} = \dfrac{-3.08}{0.15} = -21$ N. Thus, the average force on the ball is 21 N, away from the wall.

26. $v_i = +13$ m/s and $v_f = -13$ m/s. The force <u>by</u> the wall <u>on</u> the water is $3.9(-13-13) = -101$ N, or 101 N away from the wall. The force exerted by the water on the wall is the reaction force, so its magnitude is 101 N.

27. $(16)(4.8) = (16 + 4.5)v$; $v = 3.7$ m/s

28. Momentum is conserved in all explosions. Thus, the sum of the pieces' momenta is equal to the total momentum before the explosion, which is $(1500)(420) = 6.3 \times 10^5$ kg·m/s, north.

Copyright © by June Chung – All rights reserved. Reproduction of this publication without the written permission of the copyright owner is strictly prohibited.

29. $(0.12)(0.30)+(0.22)(-v_i)=(0.12)(-0.15)+(0.22)(-0.5v_i)$; $v_i = 49$ cm/s

30. First explosion: $(6.0)(0)=(2.0)(-7.5)+(4.0)v'$; $v' = 3.75$ m/s, east

Second explosion: $(4.0)(3.75)=(1.0)(0)+(3.0)v''$; $v'' = 5.0$ m/s, east

31. Collision between the bullet and the block: $(0.025)(350)+(1.5)(0)=(1.525)v'$; $v' = 5.74$ m/s

The equation of motion for the block embedded with the bullet:

$$-\mu F_N = -0.45(1.525)(9.8) = ma = (1.525)a;\quad a = -4.41 \text{ m/s}^2$$

$$v_f^2 = v_i^2 + 2ad;\quad 0 = 5.74^2 + 2(-4.41)d;\quad d = 3.7 \text{ m}$$

CHAPTER 5: WORK, ENERGY, AND POWER

1. work	2. work-energy principle (theorem)	3. gravitational potential
4. potential, kinetic	5. perfectly elastic	6. power
7. efficiency		

8. True
9. False – The work done by the friction force on a moving object is always negative.
10. True
11. True
12. True
13. False – The gravitational potential energy of an object may change even when the net work done on it is zero.
14. True
15. False – The power input to the machines must be known in order to compare the efficiencies.
16. True

17. d	18. b	19. b	20. c	21. a	22. d
23. c	24. d				

25. The amount of work must be increased by a factor of 9. Thus, 52 J of work needs to be done.

26. The change in the object's kinetic energy is -100 J, which, according to the work-energy principle, is the amount of net work on the object. Since the work done by the gravitational and normal forces is zero, the work done by the friction force is -100 J.

27. The initial gravitational potential energy of the object is 1300 J. The object's kinetic energy becomes twice its gravitational potential energy when the kinetic energy is two-thirds of the initial gravitational potential energy, or 880 J. At that point, the object's gravitational potential energy is 440 J, and the object is 5.0 m above the ground.

28. a) Find the acceleration of the object using Eq. (2.4): $a = \dfrac{8.2-5.5}{2.3} = 1.17$ m/s2

The applied force on the object is the net force: $F_{net} = ma = (9.5)(1.17) = 11$ N

b) Use Eq. (2.8) to find the distance of travel during the acceleration:

$$d = (5.5)(2.3) + \frac{1}{2}(1.17)(2.3)^2 = 15.8 \text{ m}$$

The work done by the applied force on the object is thus $W = Fd = (11)(15.8) = 180$ J

Copyright © by June Chung – All rights reserved. Reproduction of this publication without the written permission of the copyright owner is strictly prohibited.

Note: We can use the work-energy principle in Eq. (5.6) to find the work done on the object:

$$W = \frac{1}{2}(9.5)(8.2^2 - 5.5^2) = 180 \text{ J}$$

c) $P = \frac{W}{t} = \frac{180}{2.3} = 76$ W

29. The difference in elevation between the release point and the vertical position:

$$\Delta h = 0.65 - 0.65\cos 30° = 0.087 \text{ m}$$

Use the conservation of mechanical energy: $(0.34)(9.8)(0.087) = \frac{1}{2}(0.34)v_f^2$; $v_f = 1.3$ m/s

The speed of the pendulum is independent of the mass of the bob.

30. Use the conservation of mechanical energy between points B and C:

$$m(9.8)(15) + \frac{1}{2}m\,v_B^2 = \frac{1}{2}m\,(27)^2; \quad v_B = 21 \text{ m/s}$$

31. Initially the object is 1.6 m above the bottom of the incline and has 11.3 J of mechanical energy (with the ground as the reference). At the bottom of the incline, the object has 2.4 J of mechanical energy. Hence, 8.9 J of energy has been lost during the motion (possibly due to friction).

32. a) 15-kg object: $(\Delta E_P)_{15kg} = 0 - (15)(9.8)(0.66) = -97.0$ J

10-kg object: $(\Delta E_P)_{10kg} = (10)(9.8)(0.66) - 0 = 64.7$ J (The 10-kg object moves up 0.66 meters.)

Thus, $(\Delta E_P)_{total} = -32$ J.

b) $32 = (E_K)_{15kg} + (E_K)_{10kg} = \frac{1}{2}(15+10)v^2$; $v = 1.6$ m/s

c) $W_{net} = \frac{1}{2}(15)(1.6^2 - 0^2) = 19$ J

d) $W_{net} = F_{net}\,d = F_{net}(0.66)$; $F_{net} = 29.4$ N $= F_g - F_T = 15(9.8) - F_T$; $F_T = 118$ N

33. Power output by the motor = $Fv = (1350)(9.8)(0.85) = 11$ kW. Thus, the efficiency is 37%.

34. Power output by the appliances: $P_{out} = 900(0.65) = 585$ kW·h

When the average efficiency of the appliances rise to 82%, the power required by the appliances to generate 585 kW·h is $P_{in} = \frac{585}{0.82} = 713$ kW·h. Thus, Phil's electricity consumption is reduced by 187 kW·h per month. Then the amount he saves monthly is $187 \times 0.06 = \$11.20$.

35. Use the conservation of momentum to find the velocity of the 1.5-kg object after the collision:

$$(3.5)(0.50) + (1.5)(-1.2) = (3.5)(-0.52) + (1.5)v_f; \quad v_f = 1.18 \text{ m/s}$$

The sum of the kinetic energies before the collision: $\frac{1}{2}(3.5)(0.5)^2 + \frac{1}{2}(1.5)(-1.2)^2 = 1.52$ J

The sum of the kinetic energies after the collision: $\frac{1}{2}(3.5)(-0.52)^2 + \frac{1}{2}(1.5)(+1.18)^2 = 1.52$ J

Since the sum of the kinetic energies is conserved, the collision is perfectly elastic.

36. Set the positive direction as east.

Conservation of momentum: $(7.4)(6.7) + 0 = (7.4)v_1 + (3.7)v_2$; $v_2 = 13.4 - 2v_1$ ... (1)

Conservation of kinetic energy: $\frac{1}{2}(7.4)(6.7)^2 + 0 = \frac{1}{2}(7.4)v_1^2 + \frac{1}{2}(3.7)v_2^2$; $2(6.7)^2 + 0 = 2v_1^2 + v_2^2$ ... (2)

Substitute Eq. (1) into Eq. (2): $2(6.7)^2 = 2v_1^2 + (13.4 - 2v_1)^2$; $6v_1^2 - 53.6v_1 + 89.8 = 0$

Copyright © by June Chung – All rights reserved. Reproduction of this publication without the written permission of the copyright owner is strictly prohibited.

Use the quadratic formula to find v_1: $v_1 = 6.7$ or 2.23 m/s … (3)

Substitute Eq. (3) into Eq. (1) to find v_2: $v_2 = 0$ or 9.0 m/s

Obviously, v_1 and v_2 cannot be 6.7 m/s and 0 m/s, respectively (because it would mean that the 7.4-kg object has penetrated the 3.7-kg object). Thus, the 7.4-kg object moves at 2.2 m/s and the 3.7-kg object moves at 9.0 m/s after the collision.

37. Use the conservation of momentum to find the speed of the block at the impact:

$$(2.5)v + (0.023)(-314) = 0 + 0\text{;}\quad v = 2.89 \text{ m/s}$$

Since the surface is frictionless, use the conservation of mechanical energy to find H:

$$(2.5)(9.8)H + 0 = 0 + \frac{1}{2}(2.5)(2.89)^2\text{;}\quad H = 0.43 \text{ m}$$

CHAPTER 6: THERMAL PHYSICS

1. temperature	2. thermal energy	3. specific heat capacity
4. latent heat	5. vaporization	

6. False – The SI unit for temperature is Kelvin [K].
7. False – Substances at lower temperature are colder than those at higher temperature.
8. False – The greater the specific heat capacity of a substance, the harder it is to raise its temperature.
9. True
10. False – A material does absorb or give up thermal energy during a change of phase even though there is no change in temperature.
11. False – The conservation of heat principle states that the amount of heat given up by one part of a system is equal the amount of heat absorbed by the other part of the system.

12. c	13. c	14. a	15. b	16. d

17. Heat needed to turn ice at 0 °C to water at 0 °C: $Q_1 = (0.350)(3.33 \times 10^5) = 1.166 \times 10^5$ J

    Heat needed to raise the temperature of water from 0 to 100 °C: $Q_2 = (0.350)(4186)(100-0) = 1.465 \times 10^5$ J

    Heat needed to turn water at 100 °C to steam at 100 °C: $Q_3 = (0.350)(22.6 \times 10^5) = 7.91 \times 10^5$ J

    The total amount of thermal energy is $Q_1 + Q_2 + Q_3 = 10.5 \times 10^5$ J

18. $8700000 = m(4186)(30)\text{;}\ m = 69$ kg

19. Set up Eq. (6.4) and solve for m: $-(5.7)(4186)(-15) = m(4186)(33)\text{;}\ m = 2.6$ kg

20. Set up Eq. (6.4) and solve for T_f: $-(7.5)(450)(T_f - 1500) = (21)(4186)(T_f - 10)\text{;}\ T_f = 65$ °C

21. Set up Eq. (6.4) and solve for c: $-(0.222)c(46.8-111) = (0.345)(4186)(46.8-35.7)\text{;}\ c = 250$ J/kg·C°

22. Set up Eq. (6.4) and solve for T_f: $-(0.55)c(T_f - 49) = (2.3)c(T_f - 8.5)\text{;}\ T_f = 16$ °C

23. Power input = 1000 W; power output is $\frac{(1.0)(4186)(50-20)}{180} = 697$ W. Thus, the efficiency is 70%.

24. Define the system to include the cube of ice and the glass container. As no heat escapes or enters the system, all the heat lost by the glass container must be absorbed by the ice. The amount of heat

Copyright © by June Chung – All rights reserved. Reproduction of this publication without the written permission of the copyright owner is strictly prohibited.

absorbed by the ice as it turns to water is obtained from

$(0.275)(2100)(0-(-15))+(0.275)(3.33\times10^5)+(0.275)(4186)(15-0)=117.5$ kJ.

Use Eq. (6.2) to find the mass of the glass container: $117500 = m(840)(15-65)$; $m = 2.8$ kg

25. As the iron falls in the air after being dropped, it accelerates downward and gains speed. The gravitational potential energy of the iron is converted into kinetic energy while its mechanical energy is conserved. However, once it hits the surface of the water, it experiences resistance by the water. Its speed is thus reduced. At the bottom, the iron has

$$E_K = \frac{1}{2}(1.7)(0.23)^2 = 0.045 \text{ J}; \ E_P = 0 \text{ J}$$

Since the initial gravitational potential energy of the iron is $E_P = m_{iron}gh = (1.7)(9.8)(2.3) = 38.318$ J, this indicates that 38.273 J of energy has been absorbed by the system including both iron and water. Thus, we can set up the following equation:

$$Q = m_{iron}\,c_{iron}\,\Delta T + m_{water}\,c_{water}\,\Delta T$$

$$38.273 = (1.7)(450)\Delta T + (3.5)(4186)\Delta T$$

Solving for ΔT, we find that the system's temperature increases by 0.0025 Celsius degrees.

CHAPTER 7: WAVE

1. wave	2. mechanical	3. transverse
4. wavelength	5. compressions, rarefactions	6. period
7. wavelength, frequency	8. refraction	9. constructive, destructive
10. node	11. fundamental	12. Doppler effect

13. True
14. False - Sound waves are longitudinal waves.
15. True
16. True
17. True
18. False - A wave always refracts <u>away</u> from the normal when entering a fast medium from a slow medium.
19. True
20. False - Total internal reflection occurs when the angle of incidence is greater than the critical angle.
21. False - Diffraction occurs when a wave is partially blocked by an obstacle.
22. True
23. False - Standing waves are possible for longitudinal waves as well.
24. False - Standing waves are created through reflection and interference.
25. True

26. b	27. d	28. a	29. b	30. c	31. b
32. d	33. c	34. a			

35. $v = 1500 = \lambda f = (2.5)f$; $f = 600$ Hz; $T = 1/f = 0.0017$ s

Copyright © by June Chung - All rights reserved. Reproduction of this publication without the written permission of the copyright owner is strictly prohibited.

36. The frequency of the wave is 5 Hz, and its wavelength is 2.4 m. Thus, the speed of the wave is 12 m/s.

38. $\frac{v_i}{v_r} = \frac{340}{v_r} = \frac{\sin\theta_i}{\sin\theta_r} = \frac{\sin 35°}{\sin 20°}$; $v_r = 200$ m/s

39. $\frac{340}{1440} = \frac{\sin\theta_c}{\sin 90°}$; The critical angle is 14°.

Since the angle of incidence is less than the critical angle, the wave will be able to penetrate into water.

39.

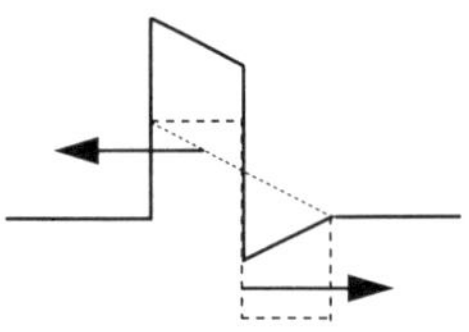

40. The frequency of this third-mode is 1416 Hz. The frequencies of the first two modes are 472 Hz and 944 Hz. Their wavelengths are 0.72 m and 0.36 m, respectively.

41. a) When two objects collide, they push each other with equal and opposite forces according to Newton's third law of motion. Their motion obeys what is predicted by the law of conservation of momentum: The sum of their momenta is conserved if there is no net external force. If we assume a perfectly inelastic collision, the objects move together after the collision, and their speed can be obtained from

$$(0.050)(+10) + (0.050)(-6.0) = (0.050 + 0.050)\vec{v}_f; \quad \vec{v}_f = +2.0 \text{ m/s}$$

b) Unlike masses colliding with each other, two energy pulses can occupy the same point in space. When the two pulses in the problem meet, they interfere constructively because their displacements are in the same direction. The resulting displacement, according to the principle of superposition, is +7.0 cm. The wavelength of each pulse remains constant during the interference. After the interference, each energy pulse continues moving in its initial direction of propagation.

CHAPTER 8: GEOMETRIC OPTICS

1. ray	2. virtual	3. focal
4. reflecting	5. lateral magnification	6. positive
7. inverted	8. bi-convex	9. real
10. lens		

11. False – An image formed by a plane mirror is as big in size as the object.
12. True
13. True
14. False – When an object is between the focal point and the vertex, a virtual image is formed.
15. True
16. True
17. True
18. False – The image characteristics are identical between images formed by a bi-convex lens and a concave mirror.

Copyright © by June Chung – All rights reserved. Reproduction of this publication without the written permission of the copyright owner is strictly prohibited.

19. False – Images formed by a diverging lens are always smaller in size than the object.

20. True

21. d	22. c	23. b	24. c	25. a

26. As shown in the diagram, the minimum length of the mirror for Randy to see the full image of himself is 86 cm.

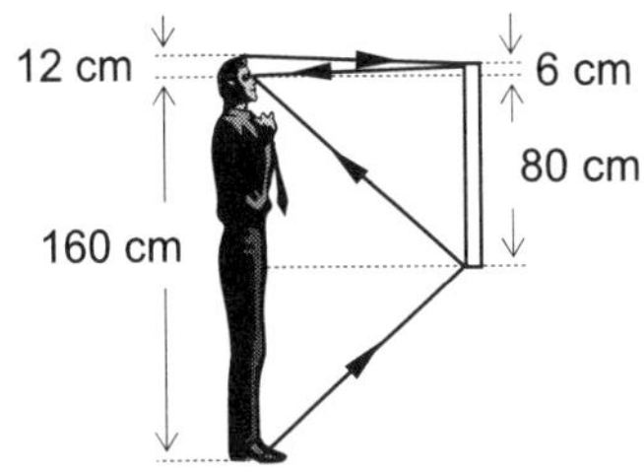

27.

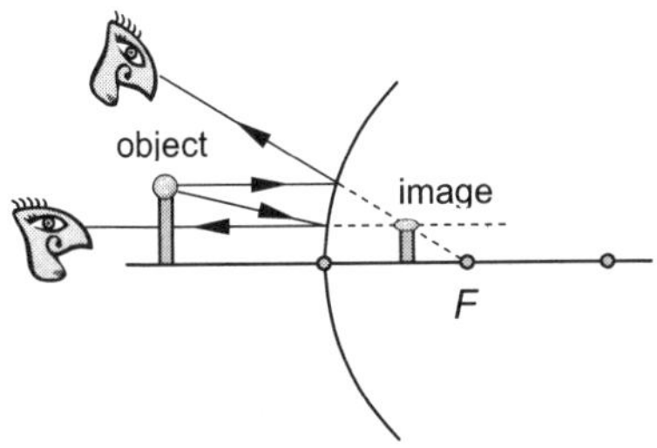

28. To find the image distance, use the mirror equation: $\frac{1}{d_o}+\frac{1}{d_i}=\frac{1}{0.073}+\frac{1}{d_i}=\frac{1}{f}=\frac{1}{-0.050}$. $d_i=-3.0$ cm, which means the image is 3.0 cm behind the mirror. The height of the image can be calculated by using Eq. (7.3): $\frac{h_i}{h_o}=\frac{0.87}{h_o}=-\frac{-0.030}{0.073}$. The object's height is 2.1 m.

29. No image is formed when an object is placed at the focal point of a concave mirror.

30. a) Since real images can be formed only by concave mirrors, the mirror is concave.

b) Use the mirror equation: $\frac{1}{d_o}+\frac{1}{d_i}=\frac{1}{25}+\frac{1}{5.0}=\frac{1}{f}$; $f=4.17$ cm

c) Use Eq. (8.2): $\frac{h_i}{40}=-\frac{5.0}{25}$; $h_i=-8.0$ cm

The image is 8.0 cm in size (it is inverted).

31. The suspect is not believable because an image formed by a convex mirror can never be inverted.

32.

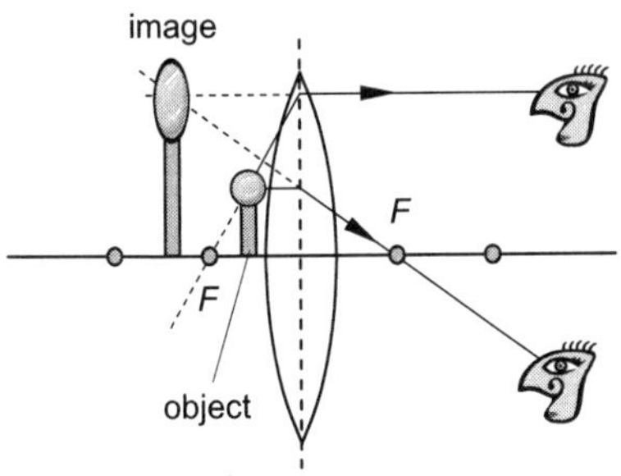

33. To find the image distance, use the lens equation: $\frac{1}{d_o}+\frac{1}{d_i}=\frac{1}{0.012}+\frac{1}{d_i}=\frac{1}{f}=\frac{1}{0.027}$. $d_i=-2.2$ cm, which means the image is 2.2 cm behind the lens (i.e., on the same side as the object). The height of the image can be calculated by using Eq. (8.4): $\frac{h_i}{h_o}=\frac{h_i}{1.8}=-\frac{-0.022}{0.012}$. The image's height is 3.2 m.

Copyright © by June Chung – All rights reserved. Reproduction of this publication without the written permission of the copyright owner is strictly prohibited.

34. Since only virtual images are formed by bi-concave lenses, $d_i = -1.2$ m. Using the lens equation, we can find the focal length of the lens: $\frac{1}{d_o} + \frac{1}{d_i} = \frac{1}{2.2} + \frac{1}{-1.2} = \frac{1}{f}$. The focal length is 2.6 m. When the object is moved to a new location, $d_i = -1.7$ m. Again using the lens equation, we find the object distance to be 4.8 m. From Eq. (8.4), $\frac{h_i}{h_o} = \frac{h_i}{0.30} = -\frac{-1.7}{4.8}$. The image height is 0.11 m, or 11 cm.

35. Since the image is virtual, d_i is negative. From Eq. (8.4), we find that the image is upright and that $d_i = -0.126$ m. Then use the lens equation to find the focal length, which is 0.29 m. The lens is a bi-concave lens because the image is upright, virtual, and smaller than the object. (The virtual images formed by a bi-convex lens are always larger than the objects.)

36. a) Use Eqs. (8.3) and (8.4). The image formed by the lens appears on the other side of the lens from the object. It is 2.29 cm from the lens and 1.40 cm in size. It is real and inverted.

    b) The image formed by the lens is 0.209 cm in front of the concave mirror. Since that image is real, we treat the rays passing through it as if they were coming from an object. Thus, the object distance and height are 0.209 cm and -1.40 cm, respectively. Using Eqs. (8.1) and (8.2), we find that the size of the final image formed by the concave mirror is 3.55 cm.

    If we use a bi-concave lens instead of a bi-convex lens, we cannot obtain a reflected image from the concave mirror because the image formed by the bi-concave lens is always virtual (which means that no light rays actually pass through the image).

CHAPTER 9: RELATIVITY

1. inertial	2. light	3. time dilation
4. length contraction	5. mass	

6. True
7. True
8. False – Michelson-Morley experiment showed that the speed of light is independent on it s direction of travel.
9. True
10. False – Two events that appear to occur simultaneously to one observer may not appear so to another observer.
11. False – An observer aboard a spaceship does not notice anything different about his watch.
12. True
13. False – The astronaut also thinks that people on Earth have gained mass.
14. False – An increase in kinetic energy at any speed results in increases in both mass and speed.

15. d	16. c	17. a	18. b	19. d	20. a

21. $t = \frac{t_o}{\sqrt{1-\left(\frac{v}{c}\right)^2}} = \frac{2.0}{\sqrt{1-\left(\frac{1.00\times10^8}{3.00\times10^8}\right)^2}} = 2.12$ s. Thus, the clock shows 07 h 00 m 02.12 s.

22. The time of travel as measured by the astronaut is the proper time. Thus,

Copyright © by June Chung – All rights reserved. Reproduction of this publication without the written permission of the copyright owner is strictly prohibited.

$$15 = \frac{12}{\sqrt{1-\left(\frac{v}{c}\right)^2}} \; ; \; v = 0.6c$$

23. $L = 0.75 = L_o\sqrt{1-\left(\frac{v}{c}\right)^2} = 1.0\sqrt{1-\left(\frac{v}{3.00\times10^8}\right)^2}$; The speed of the meter stick, v, is 2.0×10^8 m/s.

24. The length of the field measured by the people on Earth is the proper length. Thus,

    $L = 100\sqrt{1-(0.76)^2} = 65$ m

25. To observer A, observer B appears to doubled her mass. Thus, $m = 2m_o = \frac{m_o}{\sqrt{1-\left(\frac{v}{c}\right)^2}} = \frac{m_o}{\sqrt{1-\left(\frac{v}{3.00\times10^8}\right)^2}}$.

    Solving for v, we find the speed of observer A to be 2.6×10^8 m/s.

26. $\Delta E = \Delta mc^2 = 2.2(0.0001)(3.00\times10^8)^2 = 2.0\times10^{13}$ J

27. The total electricity usage of Tommy's family is 1200 kW·h, which converts to 4.32×10^9 J.

    Use Eq. (9.8) to find the required mass of the fuel:

    $\Delta E = 4.32\times10^9 = \Delta m(3.00\times10^8)^2$; $\Delta m = 4.8\times10^{-8}$ kg

    Since the conversion process is only 20% efficient, we need $5\Delta m = 2.4\times10^{-7}$ kg of fuel.

CHAPTER 10: NUCLEAR PHYSICS

1. mass number	2. strong	3. binding
4. alpha	5. beta	6. activity
7. half life	8. fission	9. critical
10. fusion		

11. False - Only protons and neutrons are nucleons.
12. False - Atomic number is determined by the number of protons inside a nucleus.
13. False - The strong force acts between all nucleons.
14. True
15. False - An electron is emitted during a beta decay.
16. False - The activity of a sample decreases exponentially with time.
17. True
18. False - Usually neutrons are used to trigger nuclear fission reactions.
19. True
20. False - Nuclear plants use nuclear fission reactions to generate electricity.

21. a	22. c	23. a	24. b	25. d

26. The atomic number for the Plutonium nucleus is 94. Thus, there are 145 neutrons inside the nucleus.

27. From Fig. 10-1, the binding energy per nucleon for a deuterium nucleus is about 1.1 MeV. The total binding energy is thus 2.2 MeV, which is 3.52×10^{-13} J. This energy corresponds to 3.9111×10^{-30} kg.

    The total mass of the nucleus is the sum of the masses of a proton and a neutron minus 3.9111×10^{-30} kg: $m = 1.6726\times^{-27} + 1.6746\times10^{-27} - 3.9111\times10^{-30} = 3.3433\times10^{-27}$ kg

Copyright © by June Chung - All rights reserved. Reproduction of this publication without the written permission of the copyright owner is strictly prohibited.

28. Alpha decay: ${}^{254}_{99}Es \rightarrow {}^{4}_{2}He + {}^{250}_{97}Bk$; Gamma decay: ${}^{250}_{97}Bk^{*} \rightarrow {}^{250}_{97}Bk + \gamma$ radiation

29. Balanced the atomic numbers: A + 0 = 79 + 1; A = 80

    Balance the mass numbers: Z + 1 = 197 + 2; Z = 198

    Thus, the identity of the unknown nucleus is mercury-198.

30. Five minutes is equivalent to four half-lifes of the substance. Hence, the number of radioactive nuclei is reduced to 1/16 of the initial value. The number of radioactive nuclei remaining after five minutes is $N = 2.4 \times 10^{16}$.

Copyright © by June Chung – All rights reserved. Reproduction of this publication without the written permission of the copyright owner is strictly prohibited.

INDEX

Copyright © by June Chung - All rights reserved. Reproduction of this publication without the written permission of the copyright owner is strictly prohibited.

Copyright © by June Chung – All rights reserved. Reproduction of this publication without the written permission of the copyright owner is strictly prohibited.

저자 약력

정 준(June Chung)

- 1996년 펜실베니아 주 카네기 멜론 대학에서 기계공학 학사 졸업
- 2002년 일리노이 대학 항공우주 공학과 박사 졸업
- 2008년 현재 캐나다 밴쿠버 소재 88 아카데미 원장

Patricia Grindlay

- 1961년 영국 맨체스터 대학에서 Bachelor of Science
- 1970년 영국 런던 대학에서 Diploma of Education 취득
- 1985년~2006년까지 밴쿠버 명문 사립고 Little Flower Academy에서 물리 및 화학 과장 역임

ACING PHYSICS 11

2009년 7월 20일 초판 인쇄
2009년 7월 25일 초판 발행

저 자 : 정 준, Patricia Grindlay
펴낸이 : 천승배
펴낸곳 : 도서출판 유한문화사

주소 : (157-801) 서울시 강서구 가양동 146-63
전화 : 2668-2055~6
팩스 : 2668-2565
http://www.yuhansa.com
E-mail : yuhansa@paran.com
등록 : 제 5-31호. 1979. 3. 6.

값 12,000 원

ISBN : 978-89-7722-117-8 93420

저자와의 협의에 의해 인지는 생략합니다.